T0299336

Elements of
Human Voice

Elements of
Human Voice

C. Julian Chen

Columbia University, USA

World Scientific

NEW JERSEY · LONDON · SINGAPORE · BEIJING · SHANGHAI · HONG KONG · TAIPEI · CHENNAI · TOKYO

Published by

World Scientific Publishing Co. Pte. Ltd.

5 Toh Tuck Link, Singapore 596224

USA office: 27 Warren Street, Suite 401-402, Hackensack, NJ 07601

UK office: 57 Shelton Street, Covent Garden, London WC2H 9HE

Library of Congress Cataloging-in-Publication Data
Names: Chen, C. Julian.
Title: Elements of human voice / C. Julian Chen, Columbia University, USA.
Description: New Jersey : World Scientific, 2016. | Includes index.
Identifiers: LCCN 2016014277 | ISBN 9789814733892 (hardcover : alk. paper)
Subjects: LCSH: Voice. | Voice--Physiological aspects. | Speech. | Sound-waves.
Classification: LCC QP306 .C44 2016 | DDC 612.7/8--dc23
LC record available at http://lccn.loc.gov/2016014277

British Library Cataloguing-in-Publication Data
A catalogue record for this book is available from the British Library.

Printed in Singapore

TO LICHING, WINSTON, KRISTIN, MARCUS, AND NORA

Preface

Voice communication through speech is a unique ability of human beings, which is probably the most significant feature differentiating humans from all other animals. In addition to speech, since the prehistory time, singing has also been a unique ability of human beings as a means to communicate with fellow human beings, and to express themselves. From the point of view of basic science, voice production is a complex physical-physiological function of human body. After the invention of telephone in 1874 and the invention of the phonograph in 1877, speech science and technology became an important field of research. Since the advance of computers in the 1940s, as the most natural means of man-machine interface, speech recognition and speech synthesis have witnessed explosive development, and are still evolving at a fast pace, as evidenced by the recent advances of Siri, Cortana, and Now. Furthermore, coded speech has always been the dominant mode in telecommunication, including wired, wireless and voice-over-IP.

Due to its importance, the science and technology of human voice has been a perennial subject of research and development since the beginnings of the scientific method and industrial revolution with Galileo and Newton. Among the enlightenment scientists, Leonhard Euler (1707–1783) occupied a unique position. Arguably the most influential mathematician of all time, Euler was also a pioneer in many fields of science, including rigid-body mechanics, fluid mechanics, astronomy, optics, acoustics, music theory, elasticity theory, civil engineering, and articulatory phonetics [47]. Among the 866 publications and communications of Euler, 10 are exclusively on acoustics. In an article on the history of acoustics, six Euler publications on acoustics are cited, more than any other author [56]. In 1727 when he was 20 years old, Euler published a treatise *Dissertatio physica de sono* (A physical dissertation on sound) [26]. It has two chapters: Propagation of Sound and Production of Sound. In Chapter 2, the physics of three categories of music instruments is discussed: string instruments, percussion and wind instruments. In paragraph 23, the physics of human voice production is discussed in analogy to the pipe organ [26]:

> Clearly the human voice is produced in the same way; indeed the epiglottis holds in place the seat of the reed tongue in the organ of speech, the vibration of which is maintained by the passage of the air ascending through the base windpipe. Besides, the vibratory motion of the air escaping from the end of the base windpipe is changed in the cavity of the mouth in a number of ways, by which the low and high-pitched tones of the voice can

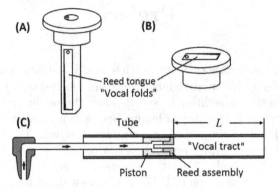

Fig. 1. Willis' mechanical devices to simulate human voice production. (A) and (B), two versions of the reed assembly. The reed tongue was made of brass, similar to those in an accordion or a harmonica. The reed tongue can vibrate freely due to a narrow gap between it and the housing. The vibration frequency of a reed tongue can be altered by its dimensions and thickness. (C), the entire experimental setup to mimic human voice production. Compressed air is blown through the entry. The timbre of different vowels is distinguished by different length L of the tube representing the vocal tract, controlled by the position of the piston. Adapted from Willis [99].

be produced, and different vocal effects are formed, which with the help of the tongue, lips, and the pharynx provide sounds with consonants.

Following Euler's conjectures, British scientist Robert Willis (1800–1875) conducted an extensive experimental and theoretical study of human voice production, and published a report *On the Vowel Sounds, and on Reed Organ-Pipes* in 1829 [99]. As a talented mechanical engineer, Willis build a number of devices to mimic human voice organs. One of those devices resembles a reed organ-pipe, where the reed tongue mimics the vocal folds, and the pipe mimics the vocal tract, see Fig 1.

Through experimental studies together with the theoretical arguments of Euler [99], Willis found that the timbre of vowel depends on the length of the tube to the right side of the piston L, which mimics the vocal tract. By changing its length, vowels [i], [e], [a], [ɔ] and [u] can be produced. The vibration frequency of the reed changes the pitch. However, the pitch has no effect on the timbre of the vowel. The timbre of the voice, which Willis called *mouth tone*, is completely independent of the pitch produced by the vocal folds, which Willis called *larynx tone*.

By quoting Euler's theoretical analysis on a transient resonator [99], see Section 1.3, Willis stated that each pulsation of air generated by the vocal folds triggers a decaying acoustic wave, the waveform of which is determined

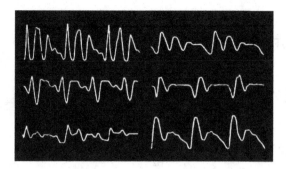

Fig. 2. Phonograph traces investigated by Ludimar Hermann. Phonograph traces of six vowels are shown [40]. Apparently, each pitch period starts with a pulsation, and then decays. For different vowels, the spectrum, or the frequency contents of the decaying elementary wave, is different. Hermann coined a term *formant* for the peak frequencies in the spectrum of the elementary waves [43].

by the vocal tract. The voice of the same vowel with different pitch is determined by the repetition rate of the pulsation of air, generated by the vocal folds. In other words, the voiced sound is produced as a superposition of elementary waves representing the timbre of the vowel, with time intervals determined by the pitch period of the vibration of vocal folds.

In 1877 Thomas Edison invented the phonograph. The waveforms of human voice can be recorded and displayed. German physiologist Ludimar Hermann (1831-1914) further amplified the mechanical grooves optically, recorded the waveforms on photographic plates, and did a large-scale quantitative study [40, 41, 42]. Some examples of the sound wave records are shown in Fig. 2. As shown, each pitch period of the waveform has a clear and consistent internal structure: Each starts with a strong pulsation, and decays within each pitch period. For a given vowel, the waveform in each pitch period is similar. Pitch is the repetition rate of the pulsations that trigger the decaying wave in each pitch period.

Ludimar Hermann observed that the spectrum of the decaying elementary wave of a vowel is peaked at a number of frequencies, characteristic of the vowel. He coined a term *formant* for those frequencies [43]. The term formant has been then used by voice and speech scientists ever since.

In modern times, waveforms of human voice can be displayed on a computer, with an accuracy far exceeding that of the phonograph. The concept of Willis becomes even more relevant. In *Elements of Acoustic Phonetics*, Peter Ladefoged (1925 – 2006) shows how the waveforms of his own voice can be straightforwardly described by Willis's concept [54].

In Chapter 7 of Ladefoged's book [54], *Production of Speech*, the vowel

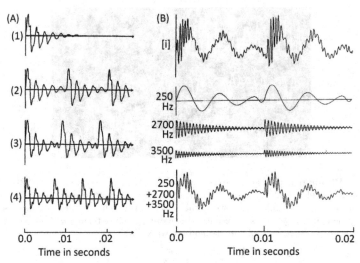

Fig. 3. Voice signals recorded and analyzed by Peter Ladefoged. (A) the vowel
[ɔ], with a dominating formant of 500 Hz. (1) triggered by a single glottal pulse. (2)
100 Hz pitch. (3) 125 Hz pitch. (4) 150 Hz pitch. Waveforms (2) through (4) can be
produced by superposing a single elementary wave with different pitch periods. (B) the
vowel [i]. It has three formants at 250 Hz, 2700 Hz and 3500 Hz [54].

[ɔ] is first discussed, see Fig. 3(A). It has a dominating formant at about
500 Hz. In example (1), with a single glottal event, the elementary wave
of the vowel [ɔ] starts with a pulsation at time 0, then decays. It can be
approximated by a decaying sinusoidal wave,

$$x(t) = a\,e^{-\kappa t}\,\sin(2\pi ft), \tag{1}$$

where a is the amplitude, κ is the decay constant, and f is the formant
frequency of the elementary wave representing the vowel.

If the pulsation repeats, as shown in Fig. 3(A) (2) through (4), vowel
sounds with different pitch frequencies are generated. In graph (2), the pitch
is 100 Hz. In graphs (3) and (4), it is 125 Hz and 150 Hz. The waveform of
the same vowel with different pitch frequencies is the *superposition* of the
decaying wave in graph (1) with different repetition rates.

Figure 3(B) shows the waveform of vowel [i] pronounced by him. Again,
for each pitch period, the waveform starts with a sharp pulsation, and then
decays. The observed waveform contains three decaying waves, starting at
the same pulsation instant, with formant frequencies 250 Hz, 2700 Hz and
3500 Hz, respectively. Ladefoged summarized thusly [54]:

> If you like to think of it in musical terms, you can say that
> corresponding to each vowel there is a chord that is characteristic

of the vowel. Owing to the pulses from the larynx, this chord is generated many times per second. There is nothing particularly new about this way of looking at speech signals. As long ago as 1829, Robert Willis said: "A given vowel is merely the rapid repetition of its particular note". This is an oversimplification, because Willis did not realize that vowels are characterized not by one frequency each but by a combination of frequencies; if we alter his remark slightly, however, and say that a given vowel is merely the rapid repetition of its peculiar chord, we have a statement that fits the data very nicely.

This way of looking at human voice should be extremely useful in speech processing, including speech recognition and speech synthesis. If one can decipher the elementary decaying waves that represent the vowels and use the spectral parameters of such elementary waves as the foundation of speech recognition, completely free from the interference of pitch, highly accurate acoustic recognition of the timbre could be achieved. On the other hand, starting from the elementary waves representing individual vowels, speech with desired prosody can be synthesized by making a superposition of those elementary waves according to the prescribed timing and intensity information. Thus synthesized speech will sound natural, because *this is exactly the way authentic human speech is produced.*

Another important issue regarding the process of human voice production is the timing between the opening and closing of the glottis and the triggering of the elementary waves. Intuitively, one would think that when the glottis is closed, the pressure from the lungs would build up. At a certain point, the air pressure in the trachea becomes high enough to break open the vocal folds and then release a sharp puff of air. That air puff is the pulsation that triggers an elementary wave [23]. However, since the invention of the electroglottograph by French physiologist Philippe Fabre in 1956 [27], a universal experimental fact was found: The speech signal is triggered by the *closing* of glottis, rather than by its opening, see Fig. 4 and Section 2.2.4. The speech signal is stronger while the glottis is closed, and weaker while the glottis is open. At the glottal opening moment, no pulsation is observed, and the speech signal decays faster. This observation directly contradicts the intuition that the air pressure in the trachea breaks open the vocal folds and then ejects a puff of air to trigger an elementary wave. Another universal experimental fact observed from the aligned voice signal and the EGG signal is that for a given hardware and software system, the sharp peak at the glottal closing moment *always has a well-defined polarity*. For databases acquired under different conditions, the polarity could either be positive, or be negative. In other words, the voice signal has a

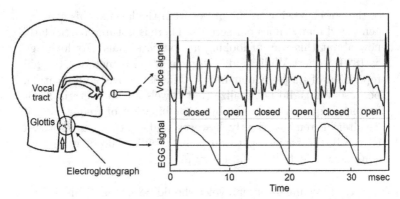

Fig. 4. Temporal correlation of voice signals and EGG signals. The starting point of an elementary wave coincides with a glottal closing moment. The glottal opening moment has little effect on the voice. See Section 2.2.4. The speech signal in the closed phase is much stronger than that in the opening phase. The time delay due to the propagation of acoustic waves from the glottis to the microphone, about 1 msec, is corrected. Adapted from Miller [62].

characteristic polarity depending on the hardware and software system of the voice data acquisition system.

The critical role of *glottal closing* in the production of voice is generally noticed. In 1992, Robert T. Sataloff made a vivid analogy of human voice production with hand clapping in an article *The Human Voice* on *Scientific American* [74], and reiterated in 2014 [76]:

> Sound is actually produced by the closing of the vocal folds, in a manner similar to the sound generated by hand clapping. Contrary to popular opinion, the vocal folds are not "cords" that vibrate like piano or guitar strings. ... (T)he more frequently they open and close, the higher the pitch.

According to that analogy, each hand clap triggers a decaying elementary acoustic wave. The superposition of such individual decaying waves constitutes a sustaining vowel sound. The pitch frequency of the vowel is the frequency of repetition of hand clapping.

The experimental fact that the elementary acoustic wave of a vowel is triggered by a *glottal closing* rather than a glottal opening is counterintuitive. To understand human voice production, it is a major mystery to be solved. In order to understand the critical role of glottal closing, the *dynamic* acoustic process inside the vocal tract immediately after a glottal closing should be studied by solving the *time-dependent wave equation*. The solution clarifies the acoustic process inside the vocal tract, and also provides a better understanding of how human voice is produced. A correct and accurate understanding of the production process of human voice enables the design of an accurate mathematical representation of voice and speech signals, which can be applied to improve speech and voice technology. And this is the synopsis of the current book.

Following is a brief summary of why and how I started to be interested in the research of physics and physiology of human voice production as well as the parameterization of voice and speech signals.

The first time I got interested in human voice was at an age of 13 as a private composition student of Professor Chéng Màojūn, the composer of the National Anthem of Republic of China. In one of his private lessons, Professor Chéng taught me that when composing vocal music based on a lyric in Mandarin, the melody line should follow the tone of the syllable to make the words clear. During my high school, college and university time, music was my primary hobby. I was a director of Peking University student philharmonic orchestra, and was frequently a conductor or a piano accompanist of choral groups. From time to time I was an arranger and composer. On the other hand, when I entered the Physics Department of Peking University, I already mastered English and Russian. Then I took French, German and Japanese in the respective departments. During the three years I lived in the center of Beijing city, I carefully listened to the native Beijing dialect speakers and made notes. By comparing with my native Shanghai dialect and several foreign languages, I wrote a 250 page monograph about a comparative study of word stresses and tone variations in Chinese languages. The manuscript was destroyed in political turmoil. However, the knowledge I accumulated was enormously helpful when I did a research on Chinese language speech recognition at IBM.

After receiving a Ph.D. in physics from Columbia University under the supervision of Professor Richard M. Osgood, I joined the Physical Sciences Department of IBM T. J. Watson Research Center as a research staff member in 1985 during the golden era of IBM's basic research. In 1986, Gerd Binnig and Heinrich Rohrer were awarded a Nobel Prize in physics for their design of the scanning tunneling microscope (STM). In 1987, Georg Bednorz and Alexander Müller were awarded another Nobel Prize in physics for their important break-through in the discovery of superconductivity in ceramic materials. I proposed an experimental project to study the mechanism of the high-temperature superconductors by building and running a low-temperature cross-sectional STM, which was fortunately approved and supported by the management. During my research work, I found that the understanding of the imaging mechanism of STM, why and how STM achieved atomic resolution, was missing. Also missing was the theory of design and calibration of the heart of STM, the tube piezoelectric scanner. I published a series of papers expounding those two problems of physics in STM, which was appreciated by the scientific community worldwide. Based on those papers, I wrote a monograph *Introduction to Scanning Tunneling Microscopy*, published by Oxford University Press in 1993.

As mentioned above, my primary academic interest has always been human voice and languages. In 1993, IBM made a decision to de-emphasize basic research in physics and to re-emphasize software technology. A high-priority research project envisioned by IBM headquarters was speech recognition for Mandarin Chinese. Based on my understanding of the tones of Mandarin Chinese, especially my comparative studies with other languages that I know, I proposed a new algorithm for its recognition. It includes a new phoneme system with tones, and a trigram language model based on mostly multisyllable *words* rather than Chinese characters [15, 16, 17, 18, 19]. My idea was diametrically opposed to the method in Mandarin speech recognition at that time. Because implementing my ideas made it possible to directly map Mandarin into an English speech recognition system, it was immediately approved by IBM's Human Language Technology Department. In the process of making the first working system, I acted as the test speaker. I also designed a statistical algorithm to automatically segment a large corpus of Chinese text into words, and made a language model from it. The first system achieved an unprecedented accuracy. Owing to the joint effort of many collaborators, the first commercial product of a Chinese language dictation system, ViaVoice Mandarin, was announced in 1997. It was then installed on almost all Chinese language computers sold in China. Because of ViaVoice Mandarin, I received an Outstanding Innovation Award from IBM. The story was documented in a science-history book [8].

In 2000, I was appointed as a technical leader for worldwide speech synthesis technology, and transferred to IBM Speech Systems Division in Boca Raton, in charge of both formant synthesizer and unit-selection synthesizer. Day after day, I stared at the waveforms of speech corpora with simultaneous electroglottograph (EGG) signals of many languages in the world. It was actually a Beagles voyage for me. At that time, the main signal processing method for unit-selection speech synthesis was PSOLA (pitch-synchronous overlap add method). As a physicist, intuition hinted that a good spectral parameterization should be pitch synchronous, rather than the pitch-asynchronous speech parameterization methods, MFCC and LPC. Intuition also hinted that by using an accurate pitch-synchronous spectral representation of human voice, a speech synthesizer with advantages of both formant synthesizer and unit-selection synthesizer could be built. However, before I retired from IBM, no practical method was found.

In January 2004, Professor Roland Wiesendanger invited me to join Department of Physics of Hamburg University, to continue research in scanning tunneling microscopy (STM). During 2004 to 2006, in Hamburg, I continued basic research in STM, and prepared the second edition of *Introduction to Scanning Tunneling Microscopy*. Although concentrated in quantum mechanics of atoms and molecules, I still kept a hobby project of singing synthesizing. Incidentally, I found that the mathematical methods in quantum mechanics fit very well to describe human voice.

In late 2006, at the invitation of Professor Osgood, I joined Department of Applied Physics and Applied Mathematics of Columbia University. To satisfy the popular demand for the science of renewable energy, I prepared a graduate-level course Physics of Solar Energy. The lectures were enthusiastically received. My lectures were video-recorded by Columbia Video Network and distributed worldwide. In 2011, a monograph and graduate-level textbook *Physics of Solar Energy* was published by John Wiley and Sons, then a Chinese translation was published in 2012.

After I returned to New York, a start-up company Voice Dream asked me to collaborate in speech synthesis technology, and provided me with software-related support. At that time, the ARCTIC speech databases from Carnegie-Melon University were publicly available. I restarted the search for a better parametrical representation of voice signals. I also met Donald Miller, an opera singer turned to human voice researcher, and found many interests in common. He pointed to me a number of unexplained experimental facts in human voice. By looking into monographs and theoretical papers on human voice from 18th century to current, I found that then popular theory of human voice production, the source-filter model, is neither the first one historically, nor the best one scientifically. A more accurate theory of human voice production could explain all experimental facts, help voice physicians and singers to improve their practice, and enable the invention of more accurate parameterizations of voice and speech signal as the foundation of new methods for voice transformation, speech recognition, speech coding, and speech synthesis [10, 11, 12, 13, 14]. In this book, I wish to share my pleasure of discovery and invention with you.

Samples of speech signals processed by the methods described in this book are posted on my homepage, www.columbia.edu/~jcc2161, under heading Human Voice. The contents are periodically updated.

The book could not be written with the help of many people. First, I thank Academician Hé Zuòxiū for recommending me for participation in an examination to enter Columbia University in 1979 despite my "right-wing-element" history. I also thank Professor T. D. Lee for choosing me as the first graduate student from the People's Republic of China to enter the Physics Department of Columbia University. His philosophy of scientific methodology, that physical intuition and order-of-magnitude estimate are far more important than mathematical development, has guided my research work for years. I sincerely thank Professor Osgood as my thesis adviser at Columbia University. His high moral standard is always my role model. I appreciate numerous discussions with Donald Miller, the founder and president of Voce Vista, on the science of singing and the understanding of human voice in general. I also want to thank Winston Chen, the founder and president of Voice Dream LLC, for giving invaluable support in software technology for my research work on human voice. I am especially grateful to Robert Sataloff for reviewing an early manuscript of this book and sending me valuable comments. I highly value Ronald Baken for his appreciation of the new concepts in this book. I am thankful to Irving P. Herman, the author of Physics of the Human Body, for helpful discussions. I sincerely appreciate the constant encouragement and support of Cevdet Noyan to my research in human voice. Finally, without the spiritual and material support of my wife Liching, the book could never have been written.

At the end of the Preface of *Introduction to Scanning Tunneling Microscopy*, I cited the following lines from the prologue of Faust by Johann Wolfgang von Goethe. My research in human voice took shape after many years of endeavor. Here those lines sound more appropriate:

Oft, wenn es erst durch Jahre durchgedrungen,
Erscheint es in vollendeter Gestalt.
Was glänzt, ist für den Augenblick geboren,
Das Echte bleibt der Nachwelt unverloren.[1]

C. Julian Chen

July 2016, Columbia University
In the City of New York

---

[1]Often, after years of perseverance, it emerges in a completed form. What glitters, is born for the moment. The Genuine lives on to the afterworld. *Faust, Vorspiel auf dem Theater.*

Contents

Part II   Mathematical Representations   95

Chapter 5:   Timbron Extraction   99

List of Figures

List of Tables

Part I

Physics and Physiology

Part I: Physics and Physiology

It can scarcely be denied that the supreme goal of all theory is to make the irreducible basic elements as simple and as few as possible without having to surrender the adequate representation of a single datum of experience.

On the Method of Theoretical Physics

Albert Einstein
The Herbert Spencer Lecture
Oxford, June 10, 1933

A good understanding of human voice production is the starting point of improving voice and developing algorithms for voice and speech technology. In Part I, a theory of human voice production is presented, which also serves as the scientific foundation of Part II, Mathematical Representations and the applications in speech and voice technology.

Chapter 1 presents background theory of acoustic waves. For simplicity, only the one-dimensional wave equation in a uniform tube is presented. It is sufficient for the understanding of the entire book.

Chapter 2 presents the basic anatomy and physiology of the voice-producing organs, including vocal folds and vocal tract. Instruments for probing and measuring the functions of the voice organs are presented. Special emphasis is directed to the non-invasive probing methods, the electroglottograph (EGG) and miniature pressure sensors, which can be applied simultaneously with the microphone during normal voicing.

Chapter 3 presents the experimental facts of human voice. First, the superposition principle formulated by Edward W. Scripture [84] is illustrated by numerous examples of voice signals. Next, the universal temporal correlation of the voice signal with the electroglottograph signal, the subglottal and the supraglottal pressures is presented. The temporal correlation strongly implies the critical role of glottal closings in voice production.

In Chapter 4, inferred from the experimental facts presented in Chapter 3, a theory of human voice production, especially for vowels, is presented. Briefly, the theory is as follows. Immediately before a glottal closure, there is a steady airflow in the vocal tract. A glottal closing abruptly blocks the airflow from the trachea into the vocal tract, triggers a zero-particle-velocity d'Alembert wavefront, which propagates and resonates in the vocal tract to form a decaying acoustic wave. Kinetic energy of the airflow in the vocal tract immediately before a closure is converted into acoustic energy. Linear superposition of these elementary resonance waves constitutes voice. The acoustic wave in the vocal tract triggered by a glottal closing is determined by the geometry of the vocal tract at that moment, thus representing the instantaneous timbre. It is reasonable to term the decaying acoustic wave trigger by a glottal closing a "timbron". The timbrons are literally the *elements* of human voice. The production mechanism of consonants is then presented, which is relatively straightforward.

In the history of the theory of human voice, especially for vowels, there are two schools of thought, analogous to the centuries-long controversy of the theory of light: the particle theory of Isaac Newton and the wave theory of Christiaan Huygens [31]. The first school of human voice, the transient theory or inharmonic theory, was proposed by British scientist Robert Willis (1800-1875) in 1829 [99]. Motivated by the similarity between human voice organ and pipe organ proposed by Leonhard Euler (1707-1783), Willis designed a series of mechanical models to artificially imitate human voice production. By following Euler's theoretical analysis, he showed that vowel sounds are composed of a series of decaying acoustic waves excited by pulsations emitted from the vocal folds. After the invention of phonograph by Thomas Edison in 1877, speech waveforms could be recorded and displayed. Ludimar Hermann, a German physiologist (1838-1914), using an optical amplification system to record the speech waveform on photographic plates, then verified Willis's theory with extensive data [40, 41, 42]. In 1902, American physiologist Edward Wheeler Scripture (1864-1945) published a monograph *The Elements of Experimental Phonetics* [83], systematically expounding the transient theory of human voice.

However, the early transient theories conjectured that the source of excitation is the air puff coming through the glottis after being pushed open by the pressure in the trachea. After the invention of electroglottograph by French physiologist Philippe Fabre in 1956 [27], a universal experimental fact was found: The speech signal is triggered by the *closing* of glottis, rather than by its opening. The waveform of the air puff during the open phase of glottis has little effect on the voice. In order to elucidate that experimental fact, in this book, the acoustic process inside the vocal tract immediately after a glottal closing is studied by solving the *time-dependent wave equation*. The solution, a dynamic acoustic process inside the vocal tract, is a quantitative representation of the transient sound wave.

An alternative theory of human voice production, the overtone-resonance theory or source-filter theory, was proposed in 1837 by Sir Charles Wheatstone (1802-1875) in a comment on Willis's paper [98]. Wheatstone agreed in every respect with Willis's theory, but added an alternative view in terms of overtones and resonance [73]. Wheatstone's view was elaborated by Hermann von Helmholtz (1821-1894) in *Sensation of Tone* [38].

Wheatstone and Helmholtz assumed that the vibration of vocal folds is truly periodic with a frequency f_0. A periodic function can be treated as a Fourier series, which consists of a fundamental component with frequency f_0, and the overtones with frequencies $2f_0$, $3f_0$, and so on. The vocal tract can be treated as a Helmholtz resonator with resonance frequencies F_1, F_2, F_3, and so on, which are called *formant frequencies*. (An interesting historical fact is that the term "formant" was coined by Ludmir Hermann [31].) If the frequency of an overtone nf_0 is equal or very close to a formant frequency F_m, it is reinforced. Therefore, the intensity envelope of the overtones on a frequency scale exhibits the spectrum of formants, or the resonance frequencies of the vocal tract. Experimentally, for sustained vowels with a fixed fundamental frequency, that theory is valid.

In late 19th century, the controversy between the Euler-Willis transient theory and the Wheatstone-Helmholtz overtone-resonance theory ran red hot. John William Strutt, also known as Lord Rayleigh (1842-1919), described both theories in detail in his monograph *Theory of Sound* [72]. Here are the subsection titles on human voice in the Table of Contents:

> Willis' theory of vowel sounds. Artificial imitation. Helmholtz's form of the theory. No real inconsistency. Relative pitch characteristic, versus fixed pitch characteristic. Auerbach's results. Evidence of phonograph. Hermann's conclusions. His analysis of A. Comparison of results by various writers. ...

According to Lord Rayleigh, the acoustic treatment of the subject of vowel production dated from a "remarkable memoir by Willis", and spent two full pages to quote Willis's original paper, including Euler's theoretical analysis of a transient resonator [72]. Then, after describing Wheatstone and Helmholtz's overtone-resonance theory, Lord Rayleigh argued that for an infinite array of pulsations with equal time interval, the only observable waves are the overtones of the fundamental frequency of the vocal cord vibration. In this case, the result of the Euler-Willis theory becomes identical to that of the Wheatstone-Helmholtz theory. Lord Rayleigh concluded, "From these considerations it will be seen that both ways of regarding the subject are legitimate and not inconsistent with one another."

Nevertheless, such a consistency is a one-way street. By definition, the subject matter of the overtone-resonance theory or source-filter theory is truly periodic signals over a sufficiently long stretch of time. For mechanical devices or music instruments, that condition is easy to fulfill. However, humans never produce a voice with truly periodic pitch. The pitch period, or the time interval between two consecutive glottal closures, varies constantly. Even if a person intentionally makes a voice of constant pitch, jitter (random variation of pitch periods), shimmer (random variation of intensity), and vibrato (in singing) always present. Synthesized voice without jitter, shimmer, and vibrato sounds buzzy, boring, and unnatural. In addition, in normal speech, pitch varies constantly to convey prosody. Within the time interval of a single vowel, pitch often varies by more than 6 semitones. More than 60% of the world's languages are tone languages [103], where pitch variations in individual syllables distinguish lexical or grammatical meaning. Isolated glottal closures and isolated decaying acoustic waves at the beginning or the end of a vowel are indispensable elements of speech. In speech, vocal fry is not unusual, where the pitch is lower than the average pitch and somewhat irregular. Among professional narrators and young women, vocal fry near the end of phrases is intentionally practiced to make their speech stylish and attractive. A startling fact is that the vowels in the vocal-fry sections of speech can be clearly perceived. It indicates that a single decaying acoustic wave triggered by a single glottal closure contains sufficient timbre information of the vowel. From the point of view of the transient theory, such phenomena are part of genuine human voice that can be treated naturally and straightforwardly.

On the other hand, by applying the transient theory to a truly periodic train of pulsations, all the phenomena predicted by the overtone-resonance theory can be derived. Because any experimental datum treatable by the overtone-resonance theory can also be treated equally well by the transient theory, and a substantial portion of experimental data can be treated only by the transient theory but not by the overtone-resonance theory, according to Einstein's criterion of a good theory, the transient theory, especially the concept of timbron introduced in Chapter 4, is presented as the sole irreducible basic element of voice production throughout the book.

Chapter 1

Acoustic Waves

Acoustic waves in air are the carrier of human voice. The production, transmission, and receiving of human voice follow the laws of acoustics. The theory of acoustic waves is a very mature branch of physics. There are excellent monographs and textbooks on this subject [63, 72]. In this Chapter, we will review the basic theory and facts of acoustic waves in air as the background information for the understanding of human voice.

1.1 Wave Equation in a Tube

An acoustic wave in a tube of uniform cross section is a simple case, but worthy thorough examination. Much of the conceptual understanding of the human voice can be obtained by considering this simple case. To help us understand the acoustic process of human voice production, the derivation is made as transparent as possible, and illustrated by a number of figures.

1.1.1 Particle displacement and perturbation density

Figure 1.1 shows a tube with a uniform cross section A. Air is a compressible fluid. A plane in the air column can only move in the x-direction. The displacement ξ of a plane in the air is a function of original location x of the plane, and time t. Suppose at a time t, the displacement of an air particle originally at x is $\xi(x,t)$, and the displacement of an air particle originally at $x + \Delta x$ is $\xi(x + \Delta x, t)$, the volume of the air mass originally in a volume

Fig. 1.1. Air displacement and perturbation density. A variation of air displacement $\xi(x,t)$ with x gives rise to a non-zero perturbation density $\rho(x,t)$.

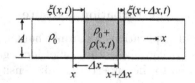

between x and $x + \Delta x$, $V = A\Delta x$, becomes

$$A\Delta x \longrightarrow A\left[(x + \Delta x + \xi(x + \Delta x, t) - (x + \xi(x, t))\right]$$
$$\approx A\Delta x \left[1 + \frac{\partial \xi(x, t)}{\partial x}\right]. \tag{1.1}$$

Assuming the mass of the air package is originally

$$m = A\Delta x \rho_0, \tag{1.2}$$

where ρ_0 is the unperturbed air density; as a consequence of conservation of mass, as the volume changes, the density is changed to

$$\frac{\rho_0 A\Delta x}{A\Delta x \left[1 + \dfrac{\partial \xi(x, t)}{\partial x}\right]} \approx \rho_0 - \rho_0 \frac{\partial \xi(x, t)}{\partial x}. \tag{1.3}$$

The additional term in the right-hand side is the *perturbation density* of the air particle, denoted as $\rho(x, t)$,

$$\rho(x, t) = -\rho_0 \frac{\partial \xi(x, t)}{\partial x}. \tag{1.4}$$

The perturbation density varies with x and t. If the particle displacement $\xi(x, t)$ increases with coordinate x, the volume of the air package is expanding, and the perturbation density is negative, as expected.

1.1.2   Particle velocity and equation of continuity

The velocity of a particle originally at location x and time t is defined as

$$u(x, t) \equiv \frac{\partial \xi(x, t)}{\partial t}. \tag{1.5}$$

By differentiating Eq. 1.4 with respect to t, using Eq. 1.5, we find

$$\frac{\partial \rho(x, t)}{\partial t} = -\rho_0 \frac{\partial u(x, t)}{\partial x}. \tag{1.6}$$

Equation 1.6 is the *equation of continuity*, which is a direct consequence of conservation of mass.

1.1.3   Perturbation pressure

In all applications regarding human voice, air can be treated as an ideal gas. According to the law of ideal gas, the unperturbed air pressure p_0 is related to the unperturbed air density ρ_0 by

$$p_0 = \rho_0 RT, \tag{1.7}$$

where R is the gas constant, 8.31 J/(mol·K), and T is the absolute temperature. Under a constant temperature, pressure is directly proportional to density. When the density is changed from ρ_0 to $\rho_0 + \rho(x,t)$, the pressure is changed from p_0 to $p_0 + p(x,t)$. The *perturbation pressure* $p(x,t)$ is

$$\frac{p(x,t)}{p_0} = \frac{\rho(x,t)}{\rho_0}. \tag{1.8}$$

When an air particle is compressed or decompressed, the temperature changes due to the work done to it. As shown in Eq. 1.1, the variation of the particle displacement $\xi(x,t)$ with x results in a change of volume,

$$\delta V = A\Delta x \frac{\partial \xi(x,t)}{\partial x}. \tag{1.9}$$

Because of the pressure p_0, an amount of work δW is done to the air particle,

$$\delta W = -p_0 \delta V = A\Delta x p_0 \frac{\partial \xi(x,t)}{\partial x} = -A\Delta x RT\rho(x,t), \tag{1.10}$$

which changes its internal energy by δW. If the specific heat of air is c_v, the heat capacity of the air particle is $c_v A\Delta x$. Temperature changes by

$$\delta T = \frac{-p_0 \delta V}{c_v A\Delta x} = \frac{RT}{c_v}\rho(x,t). \tag{1.11}$$

By differentiating a logarithmic version of Eq. 1.7, we find

$$\frac{p(x,t)}{p_0} = \frac{\rho(x,t)}{\rho_0} + \frac{\delta T}{T} = \left(1 + \frac{R}{c_v}\right)\frac{\rho(x,t)}{\rho_0}. \tag{1.12}$$

The dimensionless constant

$$\gamma = 1 + \frac{R}{c_v} \tag{1.13}$$

is a thermodynamic constant of the gas, which is the ratio between the constant-pressure specific heat and the constant-volume specific heat. For dry air, $\gamma = 1.40$. Finally we have

$$\frac{p(x,t)}{p_0} = \gamma \frac{\rho(x,t)}{\rho_0}. \tag{1.14}$$

In terms of perturbation pressure, Eq. 1.6 becomes

$$\frac{\partial p(x,t)}{\partial t} = -\gamma p_0 \frac{\partial u(x,t)}{\partial x}. \tag{1.15}$$

1.1.4   Wave equation

The wave equation can be obtained by applying Newton's law on an air particle, see Fig. 1.2. An air particle with volume $A\Delta x$ is experiencing the force by the pressure difference of the two sides. Newton's equation is

$$\rho_0 \frac{\partial u(x,t)}{\partial t} A\Delta x = -\frac{\partial p(x,t)}{\partial x} A\Delta x. \tag{1.16}$$

Differentiating both sides with regard to t, we obtain

$$\rho_0 \frac{\partial^2 u(x,t)}{\partial t^2} = -\frac{\partial^2 p(x,t)}{\partial t \partial x}. \tag{1.17}$$

On the other hand, differentiating Eq. 1.15 with regard to x yields

$$\frac{\partial^2 p(x,t)}{\partial x \partial t} = -\gamma p_0 \frac{\partial^2 u(x,t)}{\partial x^2}. \tag{1.18}$$

The differentiations should be independent of their order. Therefore, we obtain a second-order differential equation for particle velocity $u(x,t)$,

$$\rho_0 \frac{\partial^2 u(x,t)}{\partial t^2} = \gamma p_0 \frac{\partial^2 u(x,t)}{\partial x^2}. \tag{1.19}$$

Defining the *velocity of sound* c by

$$c^2 = \frac{\gamma p_0}{\rho_0} = \gamma RT, \tag{1.20}$$

we obtain the wave equation in air,

$$\frac{\partial^2 u(x,t)}{\partial x^2} = \frac{1}{c^2} \frac{\partial^2 u(x,t)}{\partial t^2}. \tag{1.21}$$

Because of the linear relation between velocity and perturbation pressure, the same type of equation is valid for the perturbation pressure,

$$\frac{\partial^2 p(x,t)}{\partial x^2} = \frac{1}{c^2} \frac{\partial^2 p(x,t)}{\partial t^2}. \tag{1.22}$$

At $0°C$ and one atmosphere pressure, the velocity of sound is $c = 330$ m/s. It is proportional to the square root of the absolute temperature. At $20°C$, $c = 342$ m/s; and at $37°C$, $c = 352$ m/s.

Fig. 1.2.   Derivation of wave equation. The wave equation is a direct consequence of Newton's equation, the conservation of mass, and the state equation of the ideal gas.

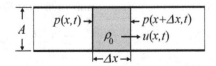

1.2   d'Alembert Solution

For an infinitely long tube of uniform cross section, Eq. 1.21 has a simple and general solution, obtained by French mathematician and physicist Jean le Rond d'Alembert in 1747,

$$u(x,t) = F(x - ct) + G(x + ct), \tag{1.23}$$

where $F(x)$ and $G(x)$ are two independent, arbitrary functions. The solution can be proved by direct verification. On one hand,

$$\frac{\partial u(x,t)}{\partial t} = c\,F'(x - ct) - c\,G'(x + ct), \tag{1.24}$$

thus

$$\frac{\partial^2 u(x,t)}{\partial t^2} = c^2\,F''(x - ct) + c^2\,G''(x + ct). \tag{1.25}$$

On the other hand,

$$\frac{\partial^2 u(x,t)}{\partial x^2} = F''(x - ct) + G''(x + ct). \tag{1.26}$$

Therefore, the d'Alembert solution satisfies the wave equation, Eq. 1.22. It is a combination of a wave $F(x - ct)$ propagating in $+x$ direction at velocity c, and a wave $G(x + ct)$ propagating in $-x$ direction at velocity $-c$.

As an application of d'Alembert solution, the reflection of a wave pulsation at the end of a tube is examined, see Fig. 1.4. First, consider the case of a rigid wall. At $x = L$, the particle velocity is always zero. Let a pulsation $F(x - ct)$ propagate from the negative side into the $+x$ direction. After the pulsation hits the wall, another wave $G(x + ct)$ propagating in the $-x$ direction to compensate the incoming wave such that at $x = L$, the velocity is always zero. The condition means

$$F(L - ct) + G(L + ct) = 0. \tag{1.27}$$

Therefore, the form of the reflected wave should be

$$G(x) = -F(2L - x). \tag{1.28}$$

Figure 1.3(A) shows the evolution of the reflection event. As shown, the wave reflected at a rigid wall has a particle velocity opposite to that of the incoming wave.

If the end of the tube is open, the perturbation pressure should always be zero. According to Eq. 1.15, the condition at $x = L$ is

$$\left.\frac{\partial u(x,t)}{\partial x}\right|_{x=L} = 0, \tag{1.29}$$

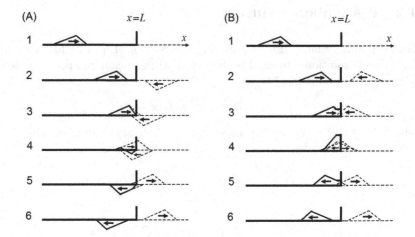

Fig. 1.3.   Reflection of acoustic waves. (A), reflection of a pulsation of particle velocity by a rigid wall at $x = L$. (1) the pulsation is propagating into the $+x$ direction. (2), the reflected wave can be represented as a d'Alembert wave propagating into the $-x$ direction. (3) through (5), the d'Alembert wave propagating into $-x$ compensates the incoming pulsation at $x = L$, such that the particle velocity is zero. (B), same, but reflected by an open end, where the perturbation pressure is always zero. The d'Alembert wave propagating into $-x$ makes the particle velocity zero at $x = L$. For a pulsation of perturbation density, (A) represents the case of an open end at $x = L$, and (B) represents the case of a rigid wall.

which means that at the open end of the tube, the derivative of velocity with respect to x is zero. It requires that

$$F'(L - ct) + G'(L + ct) = 0. \tag{1.30}$$

Following a similar argument, we have

$$G'(x) = -F'(2L - x). \tag{1.31}$$

By making an integration, notice the negative sign in x, we obtain

$$G(x) = F(2L - x). \tag{1.32}$$

As shown in Fig. 1.3(B), the particle velocity of the pulsation reflected at the open end has the same direction as the incoming pulsation.

For a wave pulsation in perturbation pressure or equivalently, perturbation density, following the same argument, similar results can be obtained. The wave pulsation reflected at a solid wall has a perturbation density of the same polarity as the incoming wave pulsation, and the wave reflected at an open end has a perturbation density of the opposite polarity as the incoming wave pulsation, see also Fig. 1.3.

1.3 Euler's Transient Resonator

In a monograph *Tentamen novae theoriar musicae*, Leonhard Euler laid down a mathematical theory of music, including the physics of many musical instruments. For woodwind instruments, he derived a formula for its pitch, or frequency. He showed that the pitch depends only on the length of the tube. For the process of sound generation, Euler wrote thusly [99]:

> If a single pulsation be excited at the bottom of a tube closed at one end, it will travel to the mouth of this tube with the velocity of sound. Here an echo of the pulsation will be formed which will run back again, be reflected from the bottom of the tube, and again present itself at the mouth where a new echo will be produced, and so on in succession till the motion is destroyed by friction and imperfect reflection. If it be a compressed pulsation that is echoed from the open end of a tube, the echo will be a rarefied one and *vice versa*, but the direction of the particle velocity will be the same. On the other hand when the reflection takes place from the stopped end, the pulsation retains its density, but the propagation from the mouth of the tube of a succession of equidistant pulsations alternatively compressed and rarefied, at intervals corresponding to the time required for the pulse to travel down the tube and back again; that is to say, a short burst of the musical note corresponding to a stopped pipe of the length in question, will be produced.

Fig. 1.4. Leonhard Euler. Swiss mathematician and physicist (1707-1783), who created much of the terminology, notations, and a large number of formulas in modern mathematics, for example, Euler's formula $e^{ix} = \cos x + i \sin x$. A special case, $e^{i\pi} + 1 = 0$, is arguably the most beautiful formula in mathematics. The constant $e = 2.71828...$ is called *Euler's number*. He also made substantial contributions to physics and engineering. His portrait was printed on a Swiss bank note, a rare honor for a scientist.

Euler's argument is shown in Fig. 1.5. Here, (1) through (3) show a pulsation of compressed air propagating from the bottom of the tube to the opening. Because at the opening of the tube, the density is a constant; the reflected pulsation is made of rarefied air, see (4) through (6). At the bottom of the tube, the particle speed is always zero. The reflected pulsation is again rarefied, see (7) through (9). The pulsation is once more reflected by the open end, to become compressed, see (10) through (12).

The period of a complete cycle T is then four times of travel time along the length L:

$$T = \frac{4L}{c}, \qquad (1.33)$$

and the frequency is

$$f = \frac{c}{4L}. \qquad (1.34)$$

The above result was given by Euler in his monograph on music.

Euler further stated that because of friction and imperfect reflection (for example, due to radiation), the intensity decays with time. Because the loss of energy per period is proportional to the available energy, the decay must be exponential. For the fundamental frequency component, the particle velocity $u(t)$ can be written in a conceptually simple form

$$u(t) = a \, \sin \frac{\pi c t}{2L} \, e^{-\kappa t}, \qquad (1.35)$$

where κ is decay constant, and a is amplitude. If the top of the tube is closed, as shown in Fig. 1.5, the first reflected pulsation is compressed.

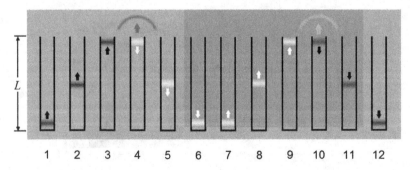

Fig. 1.5. Euler's transient resonator. (1) through (3): A pulsation of positive perturbation density propagates from the bottom of the tube to the mouth at the speed of sound. (4) through (6): In the reflected pulsation, air is rarefied. (7) through (9): The pulsation reflects from the bottom, with air rarefied, propagates to the mouth. (10) through (12): The pulsation is again reflected at the mouth to become compressed. Due to friction and imperfect reflection, the pulsation decays exponentially.

Therefore, a complete cycle is $2L/c$. The particle velocity is

$$u(t) = a \sin \frac{\pi ct}{L} e^{-\kappa t}. \tag{1.36}$$

1.4 Energy and Power

1.4.1 Power of acoustic wave

The propagation of acoustic waves carries certain energy and power with it. For an acoustic wave traveling to the $+x$ direction, acoustic energy is transferring into the $+x$ direction with a power related to the waveform. On the other hand, an acoustic wave traveling to the $-x$ direction carries acoustic energy into the $-x$ direction. In this section, the relation between the waveform and the power of acoustic wave is studied.

Consider a plane at location x, see Fig. 1.6. A perturbation pressure $p(x,t)$ is acting on the plane. The air at that location is moving with a speed $u(x,t)$. First, consider the case of a wave moving towards the $+x$ direction with air velocity $u(x,t) = F(x - ct)$. The instantaneous transfer of energy from the left-hand side to the right-hand side is

$$W = Ap(x,t)u(x,t). \tag{1.37}$$

Because both $p(x,t)$ and $u(x,t)$ are functions of $x - ct$, using Eq. 1.15, we find an expression of $p(x,t)$ in terms of $u(x,t)$ as

$$p(x,t) = \frac{\gamma p_0}{c}u(x,t). \tag{1.38}$$

The instantaneous power is

$$W = \frac{A\gamma p_0}{c}u(x,t)^2. \tag{1.39}$$

Therefore, the energy only transfers into the $+x$ direction.

Fig. 1.6.   Energy transfer of acoustic wave. For a wave traveling into the $+x$ direction, acoustic energy transfers into the $+x$ direction.

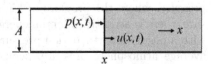

On the other hand, for a wave traveling into the $-x$ direction, where $u(x,t) = F(x + ct)$, the instantaneous power is

$$W = -\frac{A\gamma p_0}{c}u(x,t)^2. \tag{1.40}$$

The energy only transmits into the $-x$ direction, as expected.

Equations 1.36 and 1.36 are useful for estimating the instantaneous power of human voice.

1.4.2 Acoustic energy density

Another problem of interest is the acoustic energy density in a tube. The derivation is similar to the derivation of the Poynting vector and energy density in the electromagnetic field. The net instantaneous power influx into a section of unit volume is the rate of increase of energy density,

$$\begin{aligned}
\frac{d\mathcal{E}}{dt} &= -\frac{d}{dx}\left[p(x,t)u(x,t)\right] \\
&= -u(x,t)\frac{dp(x,t)}{dx} - p(x,t)\frac{du(x,t)}{dx}.
\end{aligned} \tag{1.41}$$

Using Eqs. 1.15 and 1.16,

$$\begin{aligned}
\frac{d\mathcal{E}}{dt} &= \rho_0 u(x,t)\frac{du(x,t)}{dt} + \frac{1}{\gamma p_0}p(x,t)\frac{dp(x,t)}{dt} \\
&= \frac{d}{dt}\left[\frac{\rho_0}{2}u(x,t)^2 + \frac{1}{2\gamma p_0}p(x,t)^2\right].
\end{aligned} \tag{1.42}$$

Therefore, the acoustic energy density is

$$\mathcal{E} = \frac{\rho_0}{2}u(x,t)^2 + \frac{1}{2\gamma p_0}p(x,t)^2, \tag{1.43}$$

where the first term is the acoustic kinetic energy density, and the second term is the acoustic potential energy density.

Equation 1.43 points to a significant concept in acoustic energy density. The acoustic potential energy is proportional to the square of the *perturbation pressure*. For an elementary volume with a positive perturbation pressure, where the pressure in higher than the average atmosphere pressure, there is acoustic potential energy. Similarly, for an elementary volume with a negative perturbation pressure, where the pressure is lower than the average atmosphere pressure, acoustic potential energy also exists. The magnitude of the acoustic energy only depends on the *absolute value of the perturbation pressure*, regardless of the sign of the perturbation pressure. This is similar to the case of acoustic kinetic energy. Regardless of the direction of the particle velocity, the acoustic kinetic energy is proportional to the square of local particle velocity. This point is important for the conceptual understanding of the energy conversion process, as in Section 1.5, Section 2.2.6, and Section 4.2.

1.5   Zero-Particle-Velocity Wavefronts in a Tube

In this section, a case of practical importance is studied: the conversion of aerodynamic energy into acoustic energy by blocking a steady air flow in a tube. It is important to remind that the particle velocity, the source of acoustic kinetic energy, is no different from the aerodynamic velocity of air. Both are governed by the Navier-Stokes equations and the ideal-gas state equation. Therefore, a complete conversion is possible.

1.5.1   Heuristic discussions

Figure 1.7 shows the process, step by step. For $t < 0$, shown in Fig. 1.7(A), there is a steady airflow with velocity u_0. At $t = 0$, shown in Fig. 1.7(B), a rigid wall is set up at a point $x = 0$. On both sides of the rigid wall, an acoustic process starts to take place. For the right-hand side, a d'Alembert wavefront propagates in the $+x$ direction at velocity c. Because at $x = 0$,

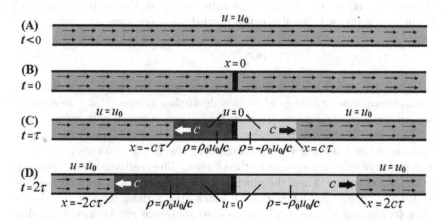

Fig. 1.7.   Zero-particle-velocity wavefronts in a tube. After the air flow is blocked at $x = 0$, two zero-particle-velocity d'Alembert wavefronts are initiated. The left one creates a compressed air column, and the right one creates a rarefied air column.

the particle velocity is always zero, the rear side of the wavefront always has zero particle velocity. Therefore, the acoustic process leaves behind a region of zero particle velocity, as shown in Fig. 1.7(C) and (D).

Because the velocity of acoustic wave and the initial particle velocity have the same direction, the air in the region left behind is rarefied, and the perturbation pressure is negative. The magnitude can be estimated as follows: As the wavefront moves to $c\tau$, where τ is a lapse of time, the continuous motion of the air with velocity u_0 makes the actual length to be $(c + u_0)\tau$. Because the conservation of mass, the density becomes

$$\rho = \rho_0 \frac{c\tau}{(c + u_0)\tau} \approx \rho_0 - \rho_0 \frac{u_0}{c}. \tag{1.44}$$

Therefore, the perturbation density is

$$\rho(x, t) = -\rho_0 \frac{u_0}{c}. \tag{1.45}$$

Using Eq. 1.14, the perturbation pressure is

$$p(x, t) = \gamma \frac{p_0 \rho(x, t)}{\rho_0} = -p_0 \frac{u_0}{c}. \tag{1.46}$$

It is negative, which stores acoustic potential energy. On the other hand, for the region $x < 0$, a d'Alembert wavefront propagates in the $-x$ direction at velocity $-c$, also as shown in Fig. 1.7(C) and (D). It also leaves behind a region of zero particle velocity. Because the velocity of acoustic wave and the initial particle velocity have opposite directions, the air in the left-over region is compressed. Similarly, the perturbation pressure is

$$p(x, t) = p_0 \frac{u_0}{c}. \tag{1.47}$$

It is positive, which also stores acoustic potential energy. Therefore, *the aerodynamic kinetic energy of the steady airflow is converted into acoustic potential energy, and generates perturbation pressures on both sides.*

The case discussed here is related to the mechanism of human voice production in the following way. Assume that the glottis is located at $x = 0$. The negative side corresponds to the trachea. The positive side corresponds to the vocal tract. When the glottis is open, there is a continuous airflow with velocity u_0. A glottal closure blocks the airflow at $x = 0$. To observe its acoustic effects, a miniature pressure sensor can be installed below the glottis, referred to as the subglottis pressure sensor; and another above the glottis, referred to as the supraglottis pressure sensor. According to the solutions presented in this section, immediately after a glottal closure, a positive perturbation pressure should be observed at the subglottis sensor, and a negative perturbation pressure of the same magnitude should be observed at the supraglottis sensor. Indeed those pressure surges were repeatedly observed, and the absolute magnitudes match well with the theoretical estimates. See Section 3.3.

Because of the importance of the process, in the following subsections, a rigorous mathematical treatment is presented.

1.5.2 Laplace-transform solution

We first formulate the problem mathematically. The initial condition at $t \le 0$ is a uniform flow of air,

$$u(x,t) = u_0, \qquad -\infty < x < \infty, \qquad t \le 0, \tag{1.48}$$

where u_0 is the velocity of air flow. At $t = 0$, a solid wall is set up at the origin, $x = 0$. After that, the velocity at $x = 0$ is always zero,

$$u(0,t) = 0, \qquad t > 0. \tag{1.49}$$

The question is to find the distribution of particle velocity $u(x,t)$, perturbation pressure $p(x,t)$ and perturbation density $\rho(x,t)$ in the entire tube as a function of x and t. To do that, the left half, $x < 0$, and the right half, $x > 0$, are treated separately using a Laplace transform in x for each half. The Laplace transform of $u(x,t)$ for $x > 0$ is,

$$U(s,t) \equiv \mathcal{L}\{u(x,t)\} = \int_0^\infty e^{-sx} u(x,t) dx. \tag{1.50}$$

The Laplace transform of the second derivative of the particle velocity $u''(x,t) = \partial^2 u(x,t)/\partial x^2$ is

$$\mathcal{L}\{u''(x,t)\} = s^2 \mathcal{L}\{u(x,t)\} - su(+0,t) - u'(+0,t). \tag{1.51}$$

Because of the boundary condition Eq. 1.1, both $u(+0,t)$ and $u'(+0,t)$ are zero for $t > 0$. Therefore, for $t > 0$,

$$\mathcal{L}\{u''(x,t)\} = s^2 \mathcal{L}\{u(x,t)\}. \tag{1.52}$$

Using wave equation Eq. 1.3, the differential equation for the Laplace transform of air velocity is

$$s^2 U(s,t) = \frac{1}{c^2} \frac{d^2 U(s,t)}{dt^2}. \tag{1.53}$$

The general solution of Eq. 1.53 is

$$U(s,t) = C_1 e^{cst} + C_2 e^{-cst}. \tag{1.54}$$

The first term in Eq. 1.54 goes to infinity for large t. Therefore, only the second term is meaningful. The constant C_2 is determined by the initial condition, Eq. 1.48,

$$C_2 = U(s,0) = \int_0^\infty e^{-sx} u_0 dx = \frac{u_0}{s}. \tag{1.55}$$

The evolution of air velocity $u(x,t)$ is determined by the inverse Laplace transform of

$$U(s,t) = \frac{u_0}{s} e^{-cst} \tag{1.56}$$

which is

$$u(x,t) = \begin{cases} 0 & : \quad 0 < x < ct \\ u_0 & : \quad x \geq ct \end{cases} \tag{1.57}$$

Equation 1.57 represents a zero-particle-velocity wave front traveling from $x = 0$ to the right at the velocity of sound c. Left over for $x < ct$ is a portion of air with zero particle velocity.

The perturbation velocity in the left-hand side is similar:

$$u(x,t) = \begin{cases} 0 & : \quad 0 > x > -ct \\ u_0 & : \quad x \leq -ct \end{cases} \tag{1.58}$$

1.5.3   Energy conversion

Before the location $x = 0$ is blocked by a solid wall, in the entire tube, there is a constant air velocity, with a uniform density of kinetic energy

$$\mathcal{E}_k = \frac{1}{2} \rho_0 u_0^2. \tag{1.59}$$

That kinetic energy is aerodynamic rather than acoustic. After the airflow is blocked at the origin $x = 0$, the velocity of the airflow in the tube gradually becomes zero. Where is that aerodynamic kinetic energy going?

As shown in Fig. 1.4(B), at time t, the zero-particle-velocity wavefront moves from $x = 0$ to $x = ct$. Because at that time, the air mass continues to flow forward with velocity u_0, it accumulates a displacement

$$\xi(ct,t) = u_0 t. \tag{1.60}$$

In the interval $0 < x < ct$, the displacement has a constant gradient

$$\frac{\partial \xi(x,t)}{\partial x} = \frac{u_0}{c}. \tag{1.61}$$

According to Eq. 1.4, in the space interval $0 < x < ct$, the perturbation density is

$$\rho(x,t) = -\frac{\rho_0 u_0}{c}, \qquad 0 < x < ct. \tag{1.62}$$

According to Eq. 1.14, the perturbation pressure is

$$p(x,t) = -\frac{\gamma p_0 u_0}{c}, \qquad 0 < x < ct. \tag{1.63}$$

The air package left behind the zero-particle-velocity wavefront is diluted and has a negative perturbation pressure. Nevertheless, according to Eq. 1.43, it gives rise to a positive *acoustic potential energy density*

$$\mathcal{E}_p = \frac{1}{2\gamma p_0} p(x,t)^2 = \frac{\gamma p_0 u_0^2}{2c^2}. \tag{1.64}$$

Notice that the velocity of sound is $c^2 = \gamma p_0/\rho_0$, Eq. 1.64 becomes

$$\mathcal{E}_p = \frac{1}{2}\rho_0 u_0^2. \tag{1.65}$$

which equals exactly the aerodynamic kinetic energy density, Eq. 1.59. In summary, the propagation of a zero-particle-velocity wavefront converts the aerodynamic kinetic energy into acoustic energy, represented by a negative perturbation pressure.

In the interval to the left of the origin, $-ct < x < 0$, the air velocity also becomes zero. A similar argument leads to the perturbation density

$$\rho(x,t) = \frac{\rho_0 u_0}{c}, \qquad -ct < x < 0. \tag{1.66}$$

According to Eq. 1.14, the perturbation pressure is

$$p(x,t) = \frac{\gamma p_0 u_0}{c}, \qquad -ct < x < 0. \tag{1.67}$$

The air left behind the left-hand side zero-particle-velocity wavefront is *compressed* with a *positive* perturbation pressure. The acoustic energy density is also identical to the original aerodynamic energy density, Eq. 1.59.

If the tube has a finite length, then the acoustic wave will radiate into open air. In this case, *the aerodynamic kinetic energy of a steady airflow is converted into acoustic energy and then radiates into open air.*

1.6 Fourier Analysis

A fundamental mathematical tool to represent and reproduce voice signals is Fourier analysis, where a voice signal can be expanded into a series of sinusoidal wave components. By using Euler's formula, $e^{ix} = \cos x + i \sin x$, the sinusoidal waves can be represented by complex exponential functions, a process that is extremely powerful and convenient. The presentation here is designed for voice signals.

1.6.1   Amplitude and phase

In 1822, French mathematician and physicist Joseph Fourier discovered that any periodic function $f(t)$ of period T satisfying the condition

$$f(t + T) = f(t) \tag{1.68}$$

can be expanded into a *fundamental frequency component* and a series of *overtones*,

$$f(t) = \sum_{n=1}^{\infty} A_n \cos\left(\frac{2n\pi t}{T} - \phi_n\right), \tag{1.69}$$

where A_1 is the amplitude and ϕ_1 is the phase of the fundamental component, whereas A_n and ϕ_n are amplitude and phase of the n-th overtone. In the conventional literature of Fourier analysis, there is also a constant, or DC component. Because of the DC-blocking capacitance in the amplifier circuit, there is no DC component in any voice signal. And the DC component is never audible. Here, the DC term is omitted.

Using a trigonometry identity, Eq. 1.69 can be rewritten into

$$f(t) = \sum_{n=1}^{\infty} \left[a_n \cos\left(\frac{2n\pi t}{T}\right) + b_n \sin\left(\frac{2n\pi t}{T}\right) \right], \tag{1.70}$$

where

$$a_n = A_n \cos \phi_n \tag{1.71}$$

and

$$b_n = A_n \sin \phi_n. \tag{1.72}$$

The phase ϕ_n can be determined by the coefficients a_n and b_n. To better define the values of the phase, the two-variable arc tangent function in C programming language is used,

$$\phi_n = \text{atan2}(b_n, a_n). \tag{1.73}$$

The returned value is in the range of $(-\pi, \pi)$. The atan2 function is well defined for every point other than $(0, 0)$, even if $x = 0$ and $y \neq 0$.

The Fourier coefficients are

$$a_m = \frac{2}{T} \int_{-T/2}^{T/2} f(\tau) \cos\left(\frac{2m\pi\tau}{T}\right) d\tau \tag{1.74}$$

and

$$b_m = \frac{2}{T} \int_{-T/2}^{T/2} f(\tau) \sin \left(\frac{2m\pi\tau}{T} \right) d\tau. \tag{1.75}$$

Those expressions of Fourier coefficients can be proved by direct computation. Substituting $f(\tau)$ by Eq. 1.70, then do the integration in Eq. 1.74. For all the terms where $n \neq m$, or any of the sine terms, the result is zero. For $n = m$, the average value of $\cos^2 x$ is $1/2$. The integral is then $T/2$, which cancels the factor $2/T$. A similar proof is valid for Eq. 1.75.

1.6.2 Complex-variable version

By using the Euler formula

$$\cos x = \frac{e^{ix} + e^{-ix}}{2}, \tag{1.76}$$

Eq. 1.69 can be written as

$$f(t) = \sum_{-\infty}^{\infty} c_n \exp \left(\frac{2n\pi it}{T} \right), \tag{1.77}$$

where we define

$$c_n = \frac{1}{2} (a_n - ib_n). \tag{1.78}$$

From Eqs. 1.74 and 1.75, $a_{-n} = a_n$ and $b_{-n} = -b_n$. Following Eqs. 1.74 and 1.75, the expression of the complex Fourier coefficients is

$$c_n = \frac{1}{T} \int_{-T/2}^{T/2} f(\tau) \exp \left(-\frac{2n\pi i\tau}{T} \right) d\tau. \tag{1.79}$$

1.6.3 Fourier transform

By increasing the period T to infinity, Fourier analysis can be applied to any function with finite values in the interval $(-\infty, \infty)$. Here is a proof.

Introducing a variable $\omega(n) = 2n\pi/T$, Eq. 1.79 becomes

$$c_n = \frac{1}{T} \int_{-T/2}^{T/2} f(\tau) e^{-i\omega(n)\tau} d\tau. \tag{1.80}$$

Substitute Eq. 1.80 into Eq. 1.77,

$$f(t) = \frac{1}{T} \sum_{n=-\infty}^{\infty} \left[\int_{-T/2}^{T/2} f(\tau) e^{-i\omega(n)\tau} d\tau \right] e^{i\omega(n)t}. \tag{1.81}$$

If T is large, the sum over n can be approximated by an integral over ω. The change of $\omega(n)$ per incremental change of n is $2\pi/T$. By writing ω as a continuous parameter, the sum in Eq. 1.81 becomes an integral,

$$f(t) \rightarrow \int_{-\infty}^{\infty} \left[\frac{1}{2\pi} \int_{-\infty}^{\infty} f(\tau) e^{-i\omega\tau} d\tau \right] e^{i\omega t} d\omega. \qquad (1.82)$$

The expression in the square bracket of Eq. 1.82 is the *Fourier transform* of the function $f(t)$,

$$F(\omega) = \frac{1}{2\pi} \int_{-\infty}^{\infty} f(\tau) e^{-i\omega\tau} d\tau, \qquad (1.83)$$

and Eq. 1.82 becomes

$$f(t) = \int_{-\infty}^{\infty} F(\omega) e^{i\omega t} d\omega. \qquad (1.84)$$

The Fourier transform of function $f(t)$ can further be represented by a pair of real functions, the amplitude spectrum $A(\omega)$ and the phase spectrum $\phi(\omega)$, defined by

$$A(\omega) = \sqrt{F^*(\omega)F(\omega)} \qquad (1.85)$$

and

$$\phi(\omega) = \text{atan2}(\text{Im}F(\omega), \text{Re}F(\omega)), \qquad (1.86)$$

where we have

$$F(\omega) = A(\omega) e^{-i\phi(\omega)}. \qquad (1.87)$$

The amplitude spectrum is a non-negative even function of ω,

$$F(-\omega) = F(\omega), \qquad (1.88)$$

and the phase spectrum is an odd function

$$\phi(-\omega) = -\phi(\omega). \qquad (1.89)$$

A single component in the Fourier series is a *sinusoidal wave*, described by the frequency or period T, the amplitude A and the phase ϕ:

$$f(t) = A \sin \left(\frac{2\pi t}{T} - \phi \right). \qquad (1.90)$$

1.7 Numerical Values

1.7.1 Pitch scale

A single sinusoidal acoustic wave has a well-defined pitch, or frequency. In music, the frequency is expressed in a logarithmic scale. The smallest interval in traditional Western music is the semitone, which is defined as a frequency ratio of $\sqrt[12]{2} = 1.059463$, see Fig. 1.8.

The numerical values of pitch p are expressed in MIDI, an integer number related to frequency f in hertz by

$$f = 440 \exp\left(\frac{\ln 2}{12}(p - 69)\right) = 440\, e^{0.05776\,(p-69)}, \qquad (1.91)$$

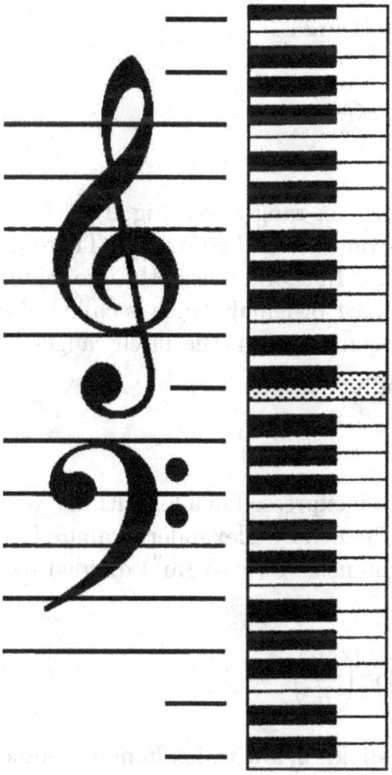

Position on the stave	Piano key	Name	MIDI	Frequency	Name	MIDI	Frequency
					(Black keys)		
		C6	84	1046.5			
		B5	83	987.77	Bb5 A#5	82	932.33
		A5	81	880.00	Ab5 G#5	80	830.61
		G5	79	783.99	Gb5 F#5	78	739.99
		F5	77	698.46			
		E5	76	659.26	Eb5 D#5	75	622.25
		D5	74	587.33	Db5 C#5	73	554.37
		C5	72	523.25			
		B4	71	493.88	Bb4 A#4	70	466.16
		A4	**69**	**440.00**	Ab4 G#4	68	415.30
		G4	67	392.00	Gb4 F#4	66	369.99
		F4	65	349.23			
		E4	64	329.63	Eb4 D#4	63	311.13
		D4	62	293.67	Db4 C#4	61	277.18
		C4	**60**	**261.63**			
		B3	59	246.94			
		A3	57	220.00	Bb3 A#3	58	233.08
		G3	55	196.00	Ab3 G#3	56	207.65
		F3	53	174.61	Gb3 F#3	54	185.00
		E3	52	164.81			
		D3	50	146.83	Eb3 D#3	51	155.56
		C3	48	130.81	Db3 C#3	49	138.59
		B2	47	123.47			
		A2	45	110.00	Bb2 A#2	46	116.54
		G2	43	97.999	Ab2 G#2	44	103.83
		F2	41	87.307	Gb2 F#2	42	92.499
		E2	40	82.407			
		D2	38	73.416	Eb2 D#2	39	77.782
		C2	36	65.406	Db2 C#2	37	69.296

Fig. 1.8. Music scale. The standard chromatic scale defined by the semitone.

Bass Baritone Tenor Contralto Mezzo Soprano

Fig. 1.9. Ranges of human voice. Usually, human voices are divided into six ranges, three female voices and three male voices. The two middle ranges are sometimes omitted, to become S, A, T, and B.

or

$$p = 69 + \frac{12}{\ln 2} \ln \frac{f}{440}. \tag{1.92}$$

Often, frequency is expressed by pitch period T, with a convenient unit of milliseconds. The relation between pitch period and pitch in MIDI is

$$p = 83.21 - 17.31 \ln T, \tag{1.93}$$

or alternatively

$$T = 122.3\, e^{-0.05776P}. \tag{1.94}$$

Usually, human voices are divided into six ranges, see Fig. 1.9. There are three female voices, soprano, mezzo-soprano, and contralto. There are three male voices, tenor, baritone, and bass. In choral music, the two middle ranges are often omitted, and the lower part of female voice is called alto, to become S, A, T, and B. In everyday conversation, the pitch ranges for both genders are often in the lower ranges, A and B.

1.7.2   Intensity scale

In engineering, the intensity of voice is often expressed in a logarithmic scale [39]. The standard unit is decibel, or dB, defined by Alexander Graham Bell, as 10 times the logarithm of the sound intensity I (in W/m^2) divided by a reference sound intensity I_0,

$$I(\text{in dB}) = 10 \log\left(\frac{I}{I_0}\right). \tag{1.95}$$

The reference sound level is hearing threshold, below which humans cannot perceive. The reference sound intensity is usually defined at 3 kHz as 10^{-12} W/m^2, to be taken as the zero point of sound level. Therefore, the sound

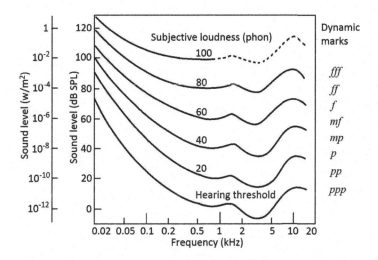

Fig. 1.10. Intensity scale of human voice. The dynamic marks on the right-hand side are heuristic rather than quantitative. Source: ISO 226:2003.

level in dB is always a positive number. Because the sound intensity is proportional to the square of sound pressure, Eq. 1.95 can be written as

$$I(\text{in dB}) = 20 \log\left(\frac{p}{p_0}\right), \tag{1.96}$$

where p is the amplitude of sound pressure, and p_0 is a reference sound pressure, which is $p_0 = 2 \times 10^{-5}$ Pa. However, the sensitivity of human ear to acoustic power depends on the frequency of the acoustic signal. Through many decades of research, the reference hearing threshold and the loudness scale is defined in an international standard, ISO 226:2003, as shown in Fig. 1.10. Human ears are most sensible to acoustic signals of frequency around 3 kHz. For acoustic signals of 200 Hz or of 10 kHz, the threshold is raised by roughly 20 dB. The dynamic marks on the right-hand side are heuristic rather than quantitative.

Chapter 2
Voice Organs

The science of human voice production involves both physics and physiology. Human voice is produced by a group of human organs. There is a vast amount of literature about the anatomy and physiology of those organs. For a brief overview, see Sataloff's *Scientific American* article *The Human Voice* [74]. More detail can be found in chapter 2 of Sundberg's *Science of the Singing Voice* [92], and first four chapters of Titze's *Principles of Voice Production* [95]. Even more detail can be found in *Clinical Anatomy and Physiology of the Voice* [75], a part of Sataloff's *Professional Voice* [78]. In this chapter, we make a brief review of the anatomy and physiology of the voice organs as a background for the discussion of human voice production mechanism in the following two chapters.

2.1 Overall Structure

Figure 2.1 is a cross-sectional view of the voice-production organs along the sagittal plane, which divides the human body into left and right halves. The source of voice energy is the kinetic energy of the airflow from the lungs through the trachea, near the bottom of Fig. 2.1. In the process of producing *voiced sounds*, including vowels such as [ɑ], [i], and [o]; and voiced consonants such as [n], [m], [z] and [ʒ]; the air stream sets the vocal folds to oscillate, causing the air path to close and open frequently. The oscillation of the vocal folds generates an alternation of finite glottal airflow and zero glottal airflow, which triggers the voiced sound. The basic anatomy of vocal folds are presented in the following section.

The "color" of the voiced sound, termed *timbre*, is determined by the vocal tract, which includes the pharynx, the oral cavity and the nasal cavity. The shape of the oral cavity can be morphed, or controlled by various parts of human organs inside the mouth, including the tongue, teeth and lips. The nasal cavity, although has no way of shaping control, is also an important path for voice generation. In the generation of nasal consonants, the oral path is closed, and the air flow can only go through the nose.

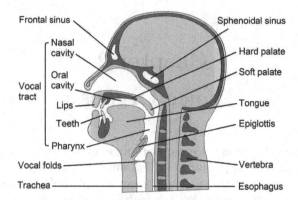

Fig. 2.1. Cross-sectional view of voice-production organs. Various parts of the figure are color-coded with different shades of gray: the bones are designated as dark gray, cartilages as medium gray, teeth as pale gray, soft tissues as light gray, and the air spaces are displayed as white. During the production of voiced sounds, the air stream through the trachea sets the vocal folds oscillating, causing the air path to close and open frequently. The timbre is controlled by the shape of the vocal tract.

The pharynx is a multiple-function organ: for food ingestion, breathing, and voice production. To prevent food from entering the trachea, the *epiglottis* functions as a gate. While swallowing food, the path between the pharynx and the vocal fold is closed by the epiglottis, the path of the pharynx and the esophagus is open, and the functions of breathing and producing voice are suspended. Inside the skull, there are many cavities filled with air, which can function as resonance devices for human voice. The largest ones are the frontal sinus and the sphenoidal sinus. The average size of the sphenoidal sinus is 2.7 cm. The average size of the sphenoidal sinus is 2.2 cm. Those cavities are connected with the nasal cavity via apertures. The walls of the sinuses are rather rigid, making them low-loss resonance cavities; and the small dimensions make the resonance frequencies high.

2.2   Vocal Folds

Vocal folds play a pivotal role in voice production. They form a valve to control the airflow from the lungs (through the trachea) to the vocal tract. While closed, vocal folds isolate the vocal tract from the porous lungs to make the vocal tract a high-quality resonance chamber. The article *An Overview of Laryngeal Function for Voice Production* by Ronald J. Baken [4] and the article *Laryngeal Function During Phonation* by Ronald C. Scherer

[80] in Sataloff's *Professional Voice* [78] cover this topic in detail. Here we present a brief review as the background for the next two chapters.

2.2.1 Anatomy

Figure 2.2 shows the top view of vocal folds from the pharynx and the cross section views for both open and closed states. The voluntary motion of the vocal folds, *abduction* or moving apart, and *adduction* or moving together, is controlled by the *vocalis muscles*. The oscillation of the vocal folds, however, is involuntary and its frequency is determined by the elastic properties and the masses of the tissue. Figure 2.2 (A) and (B) show the open or *abducted* state. The space between the separated vocal folds is the *glottis*. Through the glottis, the interior of the trachea becomes visible from the pharynx. From the cross sectional view, (B), it becomes clear that *false vocal folds* do not participate in voicing. Rather, they provide some protection to the (true) vocal folds. An important fact to notice is that the vocal folds have a finite thickness, usually greater than 5 mm. During phonation, the upper edge of the vocal folds and the lower edge of the vocal folds do not move synchronously. There is a time lag between the motion

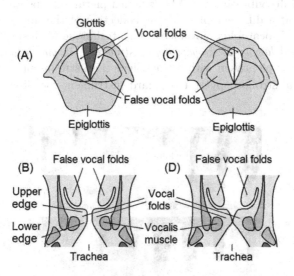

Fig. 2.2. **Vocal folds.** (A) and (C) show the top view. (B) and (D) show the cross section view along a coronal plane. Abduction of the vocal folds creates a space between them: the glottis. As shown in (B), the vocal folds have a finite thickness. During phonation, the upper edges of the vocal folds and the lower edges of the vocal folds do not move simultaneously. As shown in (D), the lower edges of the vocal folds are in contact, while the upper edges are separated, see Section Bernoulli.

of the upper edge and the motion of the lower edge. We will come back to
this fact later on. Figure 2.2 (C) and (D) show the closed or *adducted* state.
As shown in Fig 2.2(D), at that moment, the lower edges of the vocal folds
are in contact, but the upper edges are separated. There are also moments
that the upper edges of the vocal folds are in contact and the lower edges
are separated.

2.2.2 Strobovideolaryngoscopy

In principle, the motion of the vocal folds can be studied using high-speed
video recording technique. However, the vibration frequency of the vocal
fold is very high, ranging from less than 100 Hz to more than 800 Hz. To
record the motion of the vocal folds within a single pitch period, a speed
of several thousands of frames per second is required. Such systems have
become available only recently using digital technology, see Chapter 9 of
Boehme and Gross [7], but are too expensive for everyday use.

For more than fifty years, the standard technology for imaging the motion of vocal folds has been stroboscopy [7, 45]. A flash light is programmed
to produce intensive light pulses periodically, and nearly synchronous with
the periodically vibrating vocal folds. Each picture grasps an image of the
vocal folds at a different phase. The collection of the images could show
a sequence of vocal folds movement. This technique for inspecting the larynx is termed *laryngeal stroboscopy* or *strobovideolaryngoscopy*. Figure 2.3
shows an example, taken with a Kay Elemetnics stroboscope [50].

Laryngeal stroboscopy is a standard instrument for clinic diagnosis of

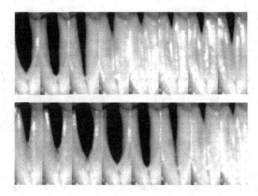

Fig. 2.3.   Strobovideolaryngoscopic images of vocal folds. the seemingly continuous series of images of the vocal folds is in fact not continuous. It is selected from a
long run of vibrations. After [50].

voice problems. However, it also has disadvantages [7, 45]. It does not provide a continuous series of images. Its time resolution is limited. It can only work with an almost strictly periodic vibration of the vocal folds, which imposes a limitation. It requires a video camera, either a rigid endoscope, or a flexible system using fiber optics inserted through the nose. In either case, it interferes with the speaker's freedom of phonation.

2.2.3 Laryngeal electromyography

The functions of the vocal folds are controlled by various muscles and nerves in the larynx. The actions of a muscle are accompanied by a change of ion concentrations in the muscles, which is then manifested as a change of electric potential in the muscle. Therefore, by continuously measuring the electric potential of a muscle, the activities of the muscle can be monitored. The technique, *electromyography*, has been a standard clinic procedure for muscle and nerve systems for several decades. This technique has been used as a clinic examination tool for the larynx, the *laryngeal electromyography* [77, 79]. In a certain sense, this technique is similar to the application of electrocardiogram for the clinical examination of the heart. There are differences, however. The actions of heart muscles are involuntary, while the actions of the muscles in the larynx is voluntary. The electric pulses of the heart are strong. Therefore, for a cardiogram, the signals can be detected by applying surface electrodes at various places on the surface of the body. Because the laryngeal muscles are small and in close proximity to each other, electrodes must be placed inside the specific muscles to effectively detect the action potentials [77].

Figure 2.4 shows a typical electrode for laryngeal electromyography. It consists of a hollow steel shaft B with a wire A runs through its center, and is insulated for the entire length except a the tip. The outer shaft B is grounded. To perform laryngeal electromyography, the patient is lying on an operating bed, with the neck extended to facilitate the identification of laryngeal muscles. Because local anesthesia could interfere with the muscle actions, and the procedure is not too painful, the needle electrodes are

Fig. 2.4. Needle electrode for laryngeal electromyography. A typical concentric needle electrode contains a central wire, insulated and enclosed in a hollow steel shaft. During examination, a needle electrode is inserted into a laryngeal muscle of interest. The outer shaft is grounded as the reference of the electric potential. After Sataloff et al. [77, 79].

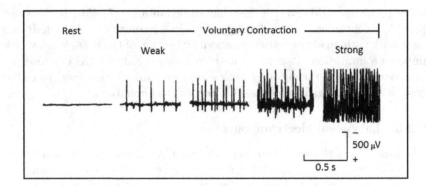

Fig. 2.5. An example of laryngeal electromyograph. When the muscle is at rest, no electric pulses are observed. The frequency of the electrical pulses, each is called a motor unit, increases as the intensity of voluntary contraction of the muscle increases. After Sataloff et al. [77, 79].

usually inserted into the muscles of interest without anesthesia. Because the actions of the laryngeal muscles are voluntary, to verify the accuracy of electrode placement, the patient is asked to take actions, for example, by sniffing or phonating the sound /i/, and the physician watches the changes in the electrical signals as evidence.

An example of the observed electromyograph is shown in Fig. 2.5. As shown, when the muscle is at rest, no electrical pulses are observed. As the intensity of voluntary contraction increases, the frequency of electrical pulses is likewise increased. In combination with other clinic observations, laryngeal electromyography can provide invaluable evidence for an accurate diagnosis of laryngeal conditions [77, 79].

The insertion of needle electrodes to the laryngeal muscles is a minimally invasive medical procedure and must be performed by highly qualified physicians. It is not performed under normal speaking or singing conditions. Therefore, other examination techniques to study the vibrations of vocal folds under normal speaking or singing conditions are needed [45]. In the following, several non-invasive techniques for the study of vocal-folds vibrations are discussed, together with the consequent findings.

2.2.4 Electroglottograpy

In 1956, French otolaryngologist Philippe Fabre, then a correspondent member of the French National Academy of Medicine, invented an electrical instrument to study the physiology of the vocal folds. The first report, entitled *Un procédé électrique percutané d'inscription de l'acclement glot-*

tique au cours de la phonetion: glottographie de haute fréquence. Premiers résultats, was published in 1957 on *Bulletin de l'Académie nationale de médecine* [27]. A schematics of its working principle is shown in Fig. 2.6. Two electrodes are pressed on the skin of the neck near the vocal folds. A weak electrical signal of frequency about 200 kHz is applied on one electrode. Another electrode is the detector of the current passing through the neck. Part of the current passes through the vocal folds. As shown in Fig. 2.6, when the glottis is open, part of the electrical current is blocked; and when the glottis is closed, more electrical current can pass through it. The small difference of conductance due to the opening and closing of the vocal folds is then detected. It is of great advantage to use an electrical signal of a high frequency: First, by using a narrow-range band-pass filter and especially by using the phase-lock technique, the signal-to-noise ratio can be greatly enhanced, and the signal quality can be made very high. Second, the temporal resolution of the signal is high. Using a 200 kHz signal, a temporal resolution of 0.02 msec can be achieved.

A follow-up paper, entitled *La Glottographie électrique en haute fréquence, particuralités de l'appareillage*, published on *Comptes Rendus, Sociéte de Biologie* in 1959 [28], disclosed a detailed electrical circuit diagram. The core of the circuit is an oscillator based on an LC circuit to produce high-frequency source. The basic circuit has been used in many contemporary products, although the vacuum tubes are replaced by transistors and integrated circuits. The commonly used name of the instrument, electroglotto-graph, abbreviated EGG, was also established. Figure 2.7 is a photograph of the electrodes, taken in the author's recording studio.

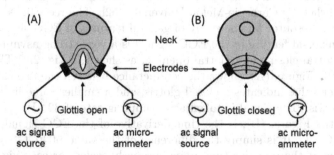

Fig. 2.6. Principle of the electroglottograph. Two electrodes are pressed against the neck, near the vocal folds. A high-frequency electrical signal, typically 100 to 200 kHz, is used to probe the electrical conductance between the two electrodes, thus to probe the status of the vocal folds. (A) While the glottis is open, the conductance between the two electrodes is lower. (B) while the glottis is closed, the conductance is higher, thus the current is higher.

**Fig. 2.7.   Electrodes of the
electroglottograph** The outer
diameter of each electrode is about
3 cm.   The circumference is the
ground connection.   The center
conductor is the signal electrode.
Photo taken in the author's recording studio.

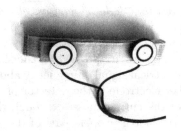

In 1992, Baken published a review article about EGG [3]. He said,

> EGG is noninvasive, innocuous, inexpensive, and does not inter-
> fere with simultaneous measurement of other relevant variables,
> such as airflow or glottal area. ... (A)nd of paramount im-
> portance, EGG provides a view of certain aspects of vocal fold
> function that is not obtainable by any other means.

In clinical voice laboratories, EGG is used routinely. It allows objective
determination of the presence or absence of glottal vibration. It reflects the
glottal condition more accurately during the closed phase, and quantitative
interpretation of the glottal condition is possible [45].

Since the 1980s, in the speech technology community, a huge amount of
high-quality simultaneous voice and EGG signals has been recorded. Designed for research and supported by the government, large-size speech and
EGG signal corpus are available publicly. The examples shown in this Section are from the ARCTIC databases, spoken by three speakers, two male
and one female, each reads a prepared text of 1132 sentences. The databases
are published by Carnegie-Melon University [52]. The recordings also provide a rich resource for the study of general features of EGG signals.

A universal feature of the EGG signals is a very strong asymmetry between glottal closing and glottal opening, as shown in Fig. 2.8. The upper
half of the Figure, (A), shows the conductance between the two electrodes.
A greater value indicates a closed glottis, and a smaller value indicates an
open glottis. The signal is taken at a sampling rate of 32 kHz. The lower
half of the Figure, (B), is the time-derivative of the EGG signal, marked
as dEGG/dt. It is simply the difference of the value of EGG signal and
the value at the previous time point, properly scaled for easy viewing. In
Fig. 2.8, the EGG signals in a pitch period are marked by numbers 1 through
8. Point 1 has the lowest conductance, the state of a widely open glottis. At
point 2, the vocal folds start to move together, and the conductance is increasing. At point 3, the conductance suddenly increases, and the dEGG/dt
signal shows a sharp maximum. At point 4, the rapid increase of conduc-

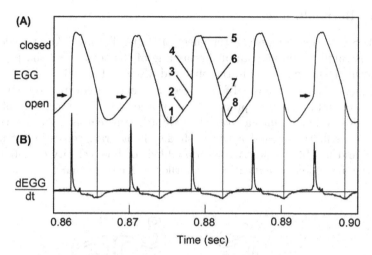

Fig. 2.8. Typical EGG signals of vowels. (A), the conductance signal. A high conductance indicates a large contact area between the two halves of the vocal folds, therefore a closed glottis. A low conductance indicates a small contact area between the two halves of the vocal folds, therefore an open glottis. (B) the time-derivative of the EGG signal, indicating the speed of conductance variation. Note the sharp peak in dEGG/dt at the closing moments. Source: ARCTIC databases, speaker bdl (US English male speaker), sentence a0002, from 0.86 second to 0.90 second, part of vowel [ɑ].

tance starts to slow down slightly, but continues to rise to a highest point 5. At that point, the conductance reaches a maximum, indicates that the contact area of the two halves of the vocal folds reaches a maximum. At point 6, the conductance is reduced, indicating that the contact area between the two vocal folds has decreased. At point 7, the decrease of conductance accelerates, showing a mild minimum in the dEGG/dt signal, indicating the opening of the glottis. At point 8, the conductance continues to drop, where the glottis opens widely.

The dramatic asymmetry of the huge peak in dEGG/dt at the closing moment (marked by thick arrows) and the weak peak in dEGG/dt at the opening moment (marked by thin vertical lines) is universal in a great majority of simultaneously acquired voice signals and EGG signals. In speech technology community, the sharp maximum of the dEGG/dt signal at the glottal closing moment is taken as the starting point of a pitch period.

The strong asymmetry of glottal opening and glottal closing is critical for the understanding of the voice-production mechanism, and is the basis of pitch-synchronous segmentation of voice signals. In the following Subsection, we will provide an explanation of the strong asymmetry.

2.2.5   Bernoulli force

In this Subsection, we discuss the effect of Bernoulli force when the glottal width is small, to provide an understanding of the strong and sharp peaks at glottal closing moments in the observed dEGG/dt signals.

Figure 2.9 shows the vocal-fold vibration process according to the body-cover vocal-fold structure proposed by Hirano [46, 87]. Each side of the vocal folds is divided into a *body*, shaded dark gray, consisting of relatively tight and stiff tissue; and a *cover*, shaded light gray, consisting of more pliable and flexible tissue. Arrows are added to show velocity distribution, with the arrow length indicating the airflow velocity at that location.

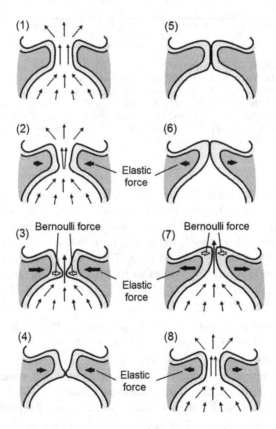

Fig. 2.9. Schematic diagram of vocal-fold vibration. After Hirano [46, 87], with the addition of velocity distributions to show the effect of the Bernoulli force. The length of the arrow indicates the velocity of airflow at that location. The numbers, (1) through (8), are made to match the numbers in Figs. 2.8 and 2.15.

At step (1), the glottis is widely open. Air from the trachea flows through the glottis. At step (2), owing to the elastic force, the vocal folds move together, and the glottal width narrows. Because of the condition of continuity, the speed of airflow through the glottis is increased. At one point, (3), Bernoulli force kicks in. Because the elastic force and the Bernoulli force are in the same direction, and the cover tissue is more pliable and flexible, a vigorous self-reinforcing cycle takes place to rapidly close the glottis, (4). The mucosal wave continues, causing the contact area between the vocal folds to increase, (5). Then, the elastic force pulls the vocal folds apart, first at the lower edges, (6). When the glottis starts to open, the Bernoulli force reappears, (7). However, this time, the Bernoulli force is in the opposite direction of the elastic force. The two forces partially cancel each other. Minor random variations of either of the Bernoulli force or the elastic will show up as the main component of force. Therefore, the opening of glottis is slow and noisy. Eventually, the elastic force overcomes the Bernoulli force. In step (8), the glottis is again widely open.

The Bernoulli force is a critical factor to make the voice production effective. Here we make an analysis to the effect of the Bernoulli force in the oscillation of the vocal folds. Based on conservation of energy, Daniel Bernoulli (1700-1782) discovered that in a continuous flow, the sum of pressure and the kinetic-energy density is a constant,

$$p_1 + \frac{1}{2}\rho v_1^2 = p_2 + \frac{1}{2}\rho v_2^2, \tag{2.1}$$

where p_1 is the pressure and v_1 is the velocity at point 1, p_2 is the pressure and v_2 is the velocity at point 2. The change of density in this case can be neglected. First, we estimate the Bernoulli pressure in the trachea. The typical rate of air flow is $Q = 0.5$ liter/sec while speaking. The typical cross section of the trachea is 2.5 cm^2 [39]. Therefore, the typical velocity is 2 m/s. The density of air is 1.25 Kg/m^3. The pressure increases by

$$\Delta p = \frac{1}{2}1.25 \times 2^2 \approx 2.5 \text{ Pa.} \tag{2.2}$$

It is tiny. However, if the glottis is nearly closed, the pressure could be large. Assuming glottis length is $L = 10$ mm [88], when the glottal width is 2 mm, the velocity is 25 m/s, and the Bernoulli pressure is about 156 times greater than that in the trachea:

$$\Delta p = \frac{1}{2}1.25 \times 25^2 \approx 400 \text{ Pa.} \tag{2.3}$$

The force is the product of Bernoulli pressure times the effective area A, which can be estimated from the parameters of the Story-Titze model [88] as

Table 2.1: Bernoulli force and elastic force

Glottal width (mm)	2	1	0.5
Bernoulli force (N)	0.006	0.024	0.096
Elastic force (N)	0.2	0.1	0.05

follows. The length of the glottis $L = 10$ mm. The total width of the vocal fold is 3 mm. Take one half of it, the effective width is 1.5 mm. Therefore, the effective area is $A = 1.5 \times 10^{-5}$ m^2. The Bernoulli force acting on the vocal folds is

$$F_B = 400 \times 1.5 \times 10^{-5} \approx 0.006 \text{ N.} \qquad (2.4)$$

Because the Bernoulli force varies as the square of velocity, it increases rapidly when glottal width is reduced, as shown in Table 2.1. Consequently, when the glottal width is small enough, the Bernoulli force becomes the dominant force to bring the vocal folds together. Once the process is started, the glottal width is further reduced, the air velocity increases further, as the Bernoulli force grows as the square of the air velocity. A self-reinforcing cycle is formed, forcefully and abruptly closing the glottis.

Since the 1960s, numerical models of vocal-fold oscillation were proposed, including the one-mass model of Flanagan and Landgraf [29], the two-mass model of Ishizaka and Flanagan [48], and the three-mass model of Story and Titze [88]. In all cases, Bernoulli force was included as a driving force together with elastic forces [29, 48, 88]. Here we present an analytic treatment to elucidate the effect of Bernoulli force near the glottal closing moment, to explain the observed sharp peaks in the dEGG/dt signals.

Figure 2.10 follows the three-mass vocal fold model of Story and Titze [88]. The three masses are the body mass m_b, the lower cover mass m_l, and the upper cover mass m_u. The values are

$m_b = 0.05$ g, $m_l = 0.01$ g, and $m_u = 0.01$ g;

and the three corresponding spring constants are

$k_b = 100$ N/m, $k_l = 5.0$ N/m, and $k_u = 3.5$ N/m.

The value of k_b is a variable. Here the medium value is taken, which corresponds to a resonance frequency

$$f = \frac{1}{2\pi}\sqrt{\frac{k_b}{m_b}} \approx 190 \text{ Hz,} \qquad (2.5)$$

a pitch frequency shared by male and female speakers and singers.

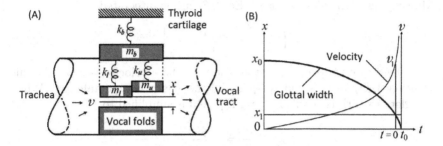

Fig. 2.10. Vocal-fold oscillation and Bernoulli force. (A) a three-mass model following Story and Titze [88]. (B) Rapid closing due to Bernoulli force. When the glottis width is small, shown as x_1, due to Bernoulli force, the speed v_1 increases rapidly. The time to close a 0.5 mm gap is much less than 0.1 msec.

Assuming the maximum glottal width is $x_0 = 2$ mm, the maximum elastic force is

$$F_k = k_b x_0 \approx 0.2\,\text{N}. \tag{2.6}$$

Comparing with the values of Bernoulli force in Table 2.1, we see that even the glottal width is at 1 mm, the Bernoulli force is still smaller than the elastic force. When the glottal width becomes 0.5 mm, at the time the elastic force is 0.05 N, the Bernoulli force becomes greater than the elastic force. Furthermore, if the glottal width is smaller than 0.5 mm, the elastic force is insignificant; and the Bernoulli force dominates.

In the following, we make a quantitative analysis of the abrupt glottal closing due to Bernoulli force, see Fig. 2.10. Because the Bernoulli force in the trachea is tiny, only the Bernoulli force of the air flow through the glottis in Eq. 2.1 is effective. Let the flow rate be Q, the anterior to posterior length of the glottis be L, the glottal width be x, the velocity of airflow is $v = Q/Lx$. When Bernoulli force dominates, shown in step (3) of Fig. 2.9, only the lower cover mass m_l is in effect. Let the effective contact area be A, the Newton's equation for the glottal width x is

$$m_l \frac{d^2x}{dt^2} = -\frac{1}{2}\rho v^2 A = -\frac{\rho Q^2 A}{2L^2}\frac{1}{x^2}. \tag{2.7}$$

Eq. 2.7 can be simplified to

$$\frac{d^2x}{dt^2} = -\frac{C}{2x^2}. \tag{2.8}$$

Taking $A = 1.5 \times 10^{-5}$ m^2, the parameter C is

$$C = \frac{\rho Q^2 A}{L^2 m_l} \approx \frac{1.25 \times (5 \times 10^{-4})^2 \times 1.5 \times 10^{-5}}{(10^{-2})^2 \times 0.01 \times 10^{-3}} \approx 6.8 \times 10^{-2}\,\frac{\text{m}^3}{\text{sec}^2}. \tag{2.9}$$

Multiplying both sides by dx/dt, Eq. 2.8 can be integrated to

$$\left(\frac{dx}{dt}\right)^2 = C\left[\frac{1}{x} - \frac{1}{x_0}\right], \tag{2.10}$$

the integration constant x_0 is the distance where velocity is zero. At this position, the glottis is widely open and the elastic force dominates. When the glottis width is small, $x \ll x_0$, Eq. 2.10 is simplified to

$$\frac{dx}{dt} = -\sqrt{\frac{C}{x}}, \tag{2.11}$$

where the minus sign from the square root indicates an attractive force. Integrating 2.11, the result can be conveniently written as

$$t = t_0 - \frac{2}{3}\sqrt{\frac{x^3}{C}}, \quad \text{with} \quad t_0 = \frac{2}{3}\sqrt{\frac{x_1^3}{C}}. \tag{2.12}$$

The meaning of Eq. 2.12 is as follows. If at $t = 0$, the glottal width is $x = x_1$; at time t_0, the glottis is completely closed, see Fig. 2.10. For $x_1 = 0.5$ mm, using the estimation of C in Eq. 2.9, one finds $t_0 \approx 0.1$ msec. On the scale of frequency, it corresponds to about 10 kHz, which is above the frequency range of formants. Therefore, on the time scale of formants, the final glottal closing event is instantaneous.

An important issue is the rate of glottal-area declination near the moment of a complete closure, that is, $x \to 0$. Because the glottal area is $\alpha = Lx$, from Eq. 2.11, the declination rate of glottal area is

$$\frac{d\alpha}{dt} = \frac{d}{dt}Lx = -L\sqrt{\frac{C}{x}}. \tag{2.13}$$

Therefore, the rate of glottal-area declination goes to infinity as the glottal width x approaches zero.

Experimental observations of glottal closing process using stroboscopy revealed that the two edges of the glottis are often not parallel, but move as a zipper with a glottic angle [7], see Fig. 2.11. By representing the glottis as a triangle with base x and angle ϕ, the velocity of airflow is $v = 2Q\phi/x^2$. Following a similar argument leading to Eq. 2.5, the Newton's equation of the glottal width x is

$$M\frac{d^2x}{dt^2} = -\frac{\rho Q^2 A\phi^2}{2x^4}. \tag{2.14}$$

Using a procedure similar to that leading to Eq. 2.10, Eq. 2.14 can be reduced to

$$\left(\frac{dx}{dt}\right)^2 = \frac{4\rho Q^2 A\phi^2}{3M}\left[\frac{1}{x^3} - \frac{1}{x_0^3}\right] \equiv D\left[\frac{1}{x^3} - \frac{1}{x_0^3}\right]. \tag{2.15}$$

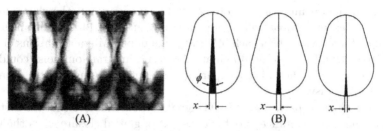

Fig. 2.11. Glottal closing with a glottic angle. (A) A sequence of images of the glottal closing process with a glottic angle, after Fig. 5.11 on Page 35 of Boehme and Gross [7]. (B) A simplified model of glottal closing with a glottic angle ϕ. The maximum glottal width x moves with the Bernoulli force. The glottal area is proportional to the square of the maximum glottal width x. Near the closing moment, the declination rate of glottal area goes to infinity.

Assuming $\phi = 0.1$, the constant D in Eq. 2.15 is

$$D \approx \frac{4 \times 1.25 \times (5 \times 10^{-4})^2 \times 1.5 \times 10^{-5} \times 0.1^2}{3 \times 0.01 \times 10^{-3}} \approx 6.25 \times 10^{-9} \frac{\text{m}^5}{\text{sec}^2}. \quad (2.16)$$

When $x \ll x_0$, the equation of motion becomes

$$\frac{dx}{dt} = -\sqrt{\frac{D}{x^3}}, \quad (2.17)$$

where the minus sign from the square root indicates an attractive force. Integrating 2.17, a result similar to Eq. 2.12 is obtained,

$$t = t_0 - \frac{2}{5}\sqrt{\frac{x^5}{D}}, \quad \text{with} \quad t_0 = \frac{2}{5}\sqrt{\frac{x_1^5}{D}}. \quad (2.18)$$

Using the value of D in Eq. 2.16, for $x_1 = 0.5$ mm, we have

$$t_0 \approx \frac{2}{5}\sqrt{\frac{(5 \times 10^{-4})^5}{6.25 \times 10^{-9}}} \approx 0.028 \text{ msec}, \quad (2.19)$$

with is even shorter than that for a parallel glottis, as expected.

Some basic conclusions are as follows:

(1) Whenever there is a complete glottal closing, the Bernoulli force will come into effect at some moment, to make the final closing explosive.

(2) Once the Bernoulli force comes into effect, the time of closing, from a finite airflow to zero, is instantaneous on the time scale of formants.

(3) From Eq. 2.11, towards the closing moment when $x \to 0$, dx/dt approaches infinity. The glottal-area declination rate goes to infinity.

(4) In the opening process of the glottis, at one point, the elastic force and the Bernoulli force will be equal in magnitude and opposite in sign. The two forces will cancel each other at some point. Minor variations of the two competing forces could dominate the event, and random noise could be clearly observed. Therefore, comparing with glottal closing, glottal opening is slow and noisy.

(5) In the interpretation of EGG signals, we should note that the instant of maximum conductance is not the instant of glottal closure. It is the moment of maximum contact area between two halves of vocal folds. Instead, the very fast variation of the EGG signal between point 3 and point 4 is the time of closing, and point 7 is the time of the glottis opening event. In other words, the peaks in the dEGG/dt charts are correlated to the closing and opening of the glottis.

2.2.6   Water-hammer analogy

The rapid closing of the glottis due to Bernoulli force points to a mechanism of human voice production proposed by Ronald Baken as an analogy to the water-hammer effect in hydrodynamics [4]:

> The sharp cutoff of flow is particularly crucial, because it is this relatively sudden stoppage of the air flow that is truly the raw material of voice. To understand why, think of an experience that you may have had with a poorly designed plumbing system. The faucet is wide open, and the water is running at full force. The tap is then quickly turned off. Water flow stops abruptly and there is a sudden THUMP! from the pipes inside the walls. (Plumbers call this "water hammer.") This happens because, in the simplest terms, the sudden cessation causes moving molecules of water to collide with those ahead of them (like the chain-reaction collision caused when a car suddenly stops on a highway). This generates a kind of "shock wave". When the pipe is jolted by this shock, it moves, creating the vibrations in the air that we hear as a thump. The relatively sudden cutoff of flow that characterizes the glottal wave creates very much the same effect in the vocal tract. An impulse-like shock wave is produced that "excites" the vibration of the air molecules in the vocal tract. That excitation is the voice in its unrefined form.

Figure 2.12 shows water-hammer effect in a water supply system with a upstream pipe, a valve, and a downstream pipe. When the valve is open, water flows continuously from the upstream pipe to the downstream pipe.

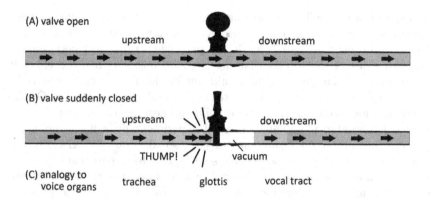

(A) valve open

upstream   downstream

(B) valve suddenly closed

upstream   downstream

THUMP!   vacuum

(C) analogy to voice organs   trachea   glottis   vocal tract

Fig. 2.12. The water-hammer effect. (A) When the valve is open, water flows continuously from upstream pipe to downstream pipe. (B) After a sudden closing of the valve, due to inertia, water continues it flow. In the upstream pipe, the flowing water collides with the valve. The kinetic energy of flowing water concentrates to the valve and creates a thump. In the downstream pipe, the continuous flow of water creates a vacuum near the valve, which could collapse the downstream pipe.

The flowing water carries a kinetic energy E_k

$$E_k = \frac{1}{2}\rho A L v^2, \tag{2.20}$$

where ρ is the density of water, A is the cross section of the pipe, L is the length of the pipe, and v is the velocity of water flow.

After a sudden closing of the valve, due to inertia, water continues to flow. In the upstream pipe, the flowing water collides with the valve. To make a numerical estimate, we use an example given by the article *water hammer* on Wikipedia as follows. In a 14 km water tunnel of 7.7 meter diameter, full of water traveling at 3.75 m/s, the kinetic energy of the flowing water is 8000 megajoules. The closing of a valve means that amount of kinetic energy must be rested. Because each kilogram of TNT corresponds to 4.7 megajoules of energy, the kinetic energy of the flowing water is equivalent to a monstrous bomb of 1.7 ton TNT. That energy can blow up a well-built water tunnel. On the downstream side, the inertia of flowing water will create a vacuum. The pipe near the valve could collapse due to the suction force created by the kinetic energy of the flowing water. To prevent pipe destruction, relief devises are often required.

Human voice production is similar to the process in the plumbing system, Fig. 2.12. The trachea is equivalent to the upstream pipe. The glottis is the valve. The vocal tract is equivalent to the downstream pipe. When the glottis is open, there is a steady airflow. When the glottis suddenly closes, a process similar to the water hammer effect in plumbing systems takes place. This time, it is air instead of water. First, air is 800 times lighter than water. Second, air is compressible. There is kinetic energy on both sides of the glottis that needs to be released, but the effect is much milder. The voice organs would not be destroyed. A *zero-particle-velocity d'Alembert wavefront* is created on each side, see Section 1.5. On the upstream side, the trachea, pressure builds up near the closed glottis. Such a pressure buildup is actually observed experimentally using a miniature pressure sensor placed inside the trachea, see Section 3.3. Because the trachea is closed, there is no audible sound. On the downstream side, the vocal tract, a zero-particle-velocity d'Alembert wavefront creates a column of rarefied air with a lower pressure. The pressure drop in the pharynx was also experimentally observed using a miniature pressure sensor placed directly above the glottis, see Section 3.3. Notice that a column of rarefied air contains the same acoustic potential energy as being compressed with the same percentage, see Section 1.4.2, it contains enough acoustic energy to become audible voice. Because the length of vocal tract is finite, the air disturbance triggered by a sudden glottal closing resonates in the vocal tract to create a decaying acoustic wave, and radiates. The waveform of that decaying acoustic wave is determined by the geometry of the vocal tract. This is the essence of human voice production, see Section 4.2.

2.2.7   Incomplete closures

The ACRTIC database has 3396 sentences [52]. A very high percentage of the EGG signals of vowels look similar to that in Fig. 2.8. However, exceptions are observed, for example the incomplete closures, shown in Figs. 2.13 and 2.14. The percentage of incomplete closures is not high, and depends on the manner of the speaker. In the ARCTIC databases, speaker bdl has 1.2%, slt has 5.35 %, and jmk has 6.13 %. Nevertheless, it is important in the understanding of the vocal-fold oscillation process.

Incomplete closures often take place near the beginning and the end of a voiced segment. At the beginning of a voiced segment, the vocal folds start to oscillate, first with a small amplitude not able to cause a full closure; then gradually increasing to make full closures. Near the end of a voiced segment, the amplitude of vocal-fold oscillation gradually decreases, the time of closing decreases, eventually becomes zero. However, even without full closure, the vocal folds continue to oscillate, as shown by the periodic variation of the EGG signals. In all cases, it is found that the vibration frequency of the vocal folds, with or without closure, is continuous. There is no abrupt change of pitch period. This experimental fact seems to indicate that the elastic force is the primary factor to determine the vibrational

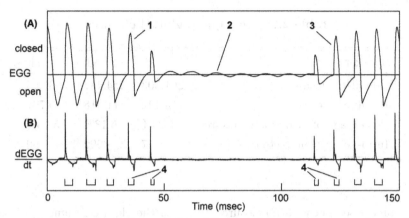

Fig. 2.13. Incomplete closures (1). (A), the EGG conductance signal. (B) the time-derivative of the EGG signal. 1, having closures, but the time of closed phase, 4, decreases gradually. 2, the vocal folds continue to vibrate without closings. 3, closure resumes with an increasing time of the closed phase 4. Source: ARCTIC databases, speaker jmk, sentence b0353, from 0.86 second to 1.01 second [52].

frequency of the vocal folds, which exists regardless of having a full closure or not.

Displayed here are several interesting cases where the closures cease for a time period then resume, often with very minor change of the vibrational frequency over the entire course. Figure 2.13 is from a male speaker of rather deep tone and strong EGG signals. At the beginning, 1, there are clear closings. Then the duration of closed phase 4 is shrinking gradually,

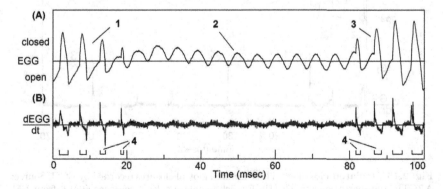

Fig. 2.14. Incomplete closures (2). Same as Fig. 2.13. Source: ARCTIC databases, speaker slt, sentence a0042, from 1.12 second to 1.22 second.

Table 2.2: Incomplete glottal closures

Speaker	bdl	jmk	slt
Complete closures	261,503	184,447	368,864
Incomplete closures	3,142	11,318	19,767
Percentage of incomplete closures	1.20%	6.13%	5.35%
Intra-voice incomplete closures	7	292	590

to enter a section with no closures, 2. Then the closure resumes with an increasing closed-phase duration, 3. Figure 2.14 is from a female speaker of bright tone. The vocal fold oscillation continues without a closure for 12 periods of steady amplitude. During the entire time of 100 msec, the change of pitch period is less than 6%. The accompanying voice signal does not change significantly with or without closures. We will return to this point in Chapter 5.

Table 2.2 shows that number and percentage of such incomplete glottal closures in the ARCTIC databases. As shown, the percentage of incomplete

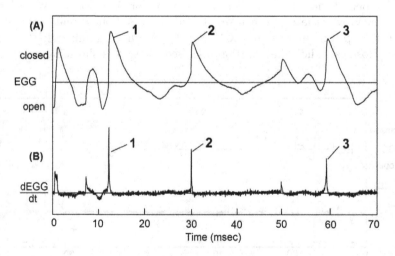

Fig. 2.15. Isolated closures. Three isolated glottal closures recorded by EGG. Source: ARCTIC databases, speaker slt (US English female speaker), sentence a0312, from 1.51 second to 1.58 second. Because the average pitch frequency is 240 Hz, the distance between adjacent closures is more than 3 times the average pitch period.

glottal closures depends on speaker, and ranges from 1% to about 6%. The number of intra-voice incomplete closures, where a section of incomplete closures is bounded by two sections with full closures, is very small.

2.2.8 Isolated closures and glottal stops

In the ARCTIC databases and the SpeechOcean databases, a substantial portion of glottal closures do not belong to a quasiperiodic series; but look like isolated events. Those isolated glottal closures are either events in a vocal-fry section, or *glottal stops*, which constitute an indispensable element of speech signals. Showing in Fig. 2.15 are three individual closures found in the recordings of a US English female speaker in the ARCTIC databases. Because the average pitch frequency is 240 Hz, the distance between adjacent closures is more than 3 times the average pitch period. The related phenomena will be discussed in Chapters 4 and 5.

2.2.9 Videokymography

In section 2.2.2, strobovideolaryngoscopy is briefly presented as a powerful method to study vocal fold oscillations. However, the time resolution of that technology is not enough to observe the fast events during phonation. The sequence of strobovideolaryngoscopic images is not a true sequence in time, but a series of sporadically chosen samples from a large number of pitch periods. Videokymography provides an alternative solution [7, 93].

The working principle of videokymography is shown in Fig. 2.16. Similar to stroboscopy, a CCD video camera is inserted in the larynx. Instead of taking full video pictures, the vertical scan is disabled, and only the horizontal scan is functioning. The line scan is fixed to a single anterior-

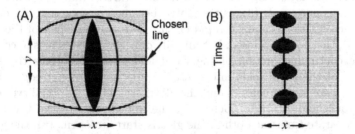

Fig. 2.16. Working principle of videokymography. (A), the video image of the vocal folds. Instead of taking full video images, the the vertical scan is disabled, and the line scan is fixed to a single anterior-posterior position. (B) An image with time as the *y* axis [7, 93].

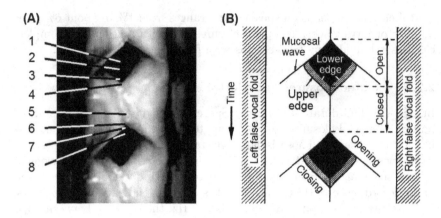

Fig. 2.17. Image obtained by videokymography. The videokymography image of a healthy person (A), with explanations in (B). The numbers, (1) through (8), are made to match the numbers in Fig. 2.8. Source: References [7, 93].

posterior position. The exact position of the line can be chosen to inspect different positions of the vocal folds, see Fig. 2.16(A). The scanning speed is increased to approximately 8000 lines per second. Excellent temporal resolution is achieved. The image, with time as the y axis, is displayed, as shown in Fig. 2.16(B). Since the vocal folds are almost white and the glottis is dark, the contrast is high.

A videokymography image of a healthy person is shown in Fig. 2.17(A), with explanation shown in Fig. 2.17(B). The numbering of points is identical to the numbering on the EGG signals in Fig. 2.8. (1), the glottis is widely open. (2), the lower edge of the vocal folds start to move together, thus the glottis is narrowed. (3), as the glottis becomes very narrow, the closing motion accelerates to make a rapid closure at the lower edges of the vocal folds. (4), the glottis is closed, but initially only the lower edge is in contact. (5), the continuing motion of the vocal fold gradually brings the entire thickness of the vocal folds into contact. (6). as the upper edges of the vocal folds are still in contact and the glottis is still closed, the lower edges of the vocal folds start to open. (7), the upper edges of the vocal folds finally separate from each other, the glottis starts to open. (8), the glottis is opened. From the images, it is clear that the closing starts at the lower edges of the vocal folds, and the opening starts at the upper edges of the vocal folds. The wave-like motion, the *mucosal wave*, is characteristic of the oscillation of the vocal folds.

2.2.10 Closed quotient and voice intensity

An important experimental observation using EGG and videokymography is the relation between the ratio of the duration of the closed phase and the pitch period with respect to voice intensity [7, 62]. The *closed quotient*, abbreviated CQ, the ratio of the duration of closed phase and the pitch period, is a significant parameter to characterize the the quality voice. Sometimes, the ratio of the duration of the open phase and the pitch period, the *open quotient*, abbreviated OQ, is also used. Obviously,

$$CQ + OQ = 1. \tag{2.21}$$

It is a general observation that high CQ is correlated to loud phonation, and low CQ is correlated to soft phonation [7, 62]. The typical experimental finding of CQ is as follows: For female voices, 0.39 for soft phonation, and 0.51 for loud phonation. For male voices, 0.35 for soft phonation, and 0.53 for loud phonation, see page 109 of Reference [7]. In the study of singing, a frequent observation is that high-intensity singing voice is correlated with high closed quotients, see pages 40–41 of Reference [62]. We will provide a quantitative explanation of the correlation between voice intensity and closed quotient in Chapter 4.

2.3 Vocal Tract

An overall diagram of the vocal tract is in Fig. 2.1. It consists of the larynx, the oral cavity, and the nasal cavity. The oral cavity can undergo extensive modifications by the motions of the soft palate, the tongue, teeth and lips. The overall dimensions are shown in Table 2.3, after Stevens [86].

Vowels are produced by combined actions of vocal folds and vocal tract. Although for vocal music, vowels are the central element, in speech, consonants sometimes hold a more important role than vowels. Evidence is

Table 2.3: Dimensions of the vocal tract

Average length in mm	Female	Male
Vocal tract length	141	169
Pharynx length	63	89
Oral cavity length	78	81

that in many writing systems in the world, only consonants are explicitly written, such as Arabic.

Fricatives are also universal. Unvoiced fricatives are generated at various places of the vocal tract. Vocal folds are widely open. No acoustic role is played by vocal folds in the production of unvoiced fricatives. As an example, fricative [s] is presented.

Since the first systematic study of human voice production by Robert Willis in 1829 [99], it is universally recognized that the timbre of vowels is determined by the shape of the vocal tract. Willis showed that for vowels, the effect of the vocal tract can be understood in terms of Euler's transient resonator. We will analyze the vowels [ɑ] and [u] as examples. More details will be presented in Chapter 3 and Chapter 4.

2.3.1   Plosives

According to Peter Ladefoged and Ian Maddieson [55], stop consonants are the most universal sound of the world's languages. Unvoiced stop consonants are generated in the front part of the oral cavity. Vocal folds are in an abducted state, functioning only as a through hole, same as at breathing. As an example, the case of plosive [k] is presented.

Compared with other sounds, the physics and physiology of the production of plosives are relatively simple. As an example, unvoiced stop [k] is discussed here. As we will show below, the actual sound depends on the vowel that follows. Do a five-second experiment by yourself and you will be convinced. Pronounce a devoiced "cut" and a devoiced "cool". Even before your make a plosive, the positions of the lips and tongue are different. The perceived vibration frequencies of two versions of [k] sound quite different. In Fig. 2.18, the case of [kʌ], such as in "cut", is shown.

The production of a velar stop [k] has two steps. In the first step, the tongue is pressing against the hard palate, and the soft palate is pressing against the back side of the pharynx, as shown in (1) of Fig. 2.18(A). The entire air path is completely blocked. Air pressure is building up in the pharynx, as shown in Fig. 2.18(A). The stop consonant starts with the voluntary release of the tongue from the hard palate, as shown by (2) of Fig. 2.18(B). A burst of air flows into the front oral cavity, as shown by (3) of Fig. 2.18(B). Because the following vowel is [ʌ], the teeth are close together, and the lips are not far from the teeth. The space between the soft palate (2) and teeth (4) forms an Euler transient resonator. An example of the observed waveform is shown by the dotted curve in Fig. 2.19. The thin solid curve is an analytic approximation of the observed data using a decaying sinusoidal wave with frequency 1.40 kHz,

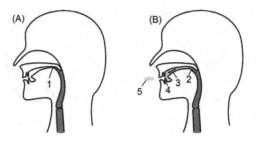

Fig. 2.18. Production of plosive [kʌ]. (A) As a preparation, the tongue presses against the hard palate to block the airflow. (B) After a voluntary release of the tongue from the hard palate, (2), air bursts into the oral cavity (3), which acts as an Euler transient resonator. A decaying acoustic wave is formed then radiates into open air, (5).

$$u(t) = 1500 \sin(2\pi \times 1.40\,t)e^{-0.485\,t}, \qquad (2.22)$$

where t is in milliseconds. The length of the Euler transient resonator can be estimated by Eq. 1.34. Using millimeters and milliseconds as units, one obtains $L = 352/(4 \times 1.40) \approx 63$ mm. It is consistent with the typical length from the back of the hard palate to the teeth for a male speaker.

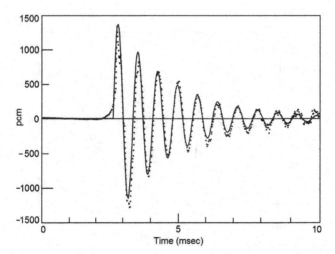

Fig. 2.19. Waveform of plosive [kʌ]. Dotted line: experimental waveform, source: SpeechOcean Database for Mandarin speech synthesis, sentence 041429, 2.557 sec to 2.567 sec. Thin solid curve is a decaying sinusoidal function, Eq. 2.22.

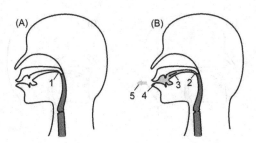

Fig. 2.20. Production of plosive [ku]. (A) As a preparation, the tongue presses against the hard palate (1) to block the airflow. (B) After a voluntary release of the tongue from the hard palate, (2), air bursts into oral cavity (3), which acts as an Euler transient resonator. A decaying acoustic wave is formed and then radiates, (5).

Another example of plosive [ku], as in "cool", is shown in Fig. 2.20. Even at the step of preparing the plosive, the contact point of the tongue against the palate is located further back, and the lips are protruded further out, making the length of the Euler transient resonator much greater than the case of [kʌ]. A recorded waveform is shown as the dotted line in Fig. 2.21. A decaying sinusoidal wave, represented by the thin solid curve, is

$$u(t) = 2400 \, \sin(2\pi \times 0.753 \, t)e^{-0.309 \, t}. \tag{2.23}$$

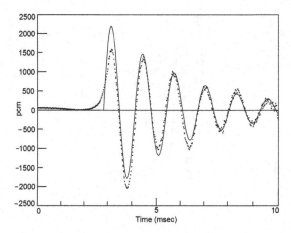

Fig. 2.21. Waveform of plosive [ku]. Dotted line: experimental waveform, source: SpeechOcean Database for Mandarin speech synthesis, sentence 004050, 3.213 sec to 3.223 sec. Thin solid curve is a decaying sinusoidal wave represented by Eq. 2.23.

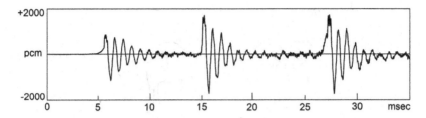

Fig. 2.22. Three resonance waves in a single plosive. (A) As a preparation, the tongue presses against the hard palate to block the airflow. (B) After a voluntary release of the tongue from the hard palate, (2), a burst of air Source: SpeechOcean Mandarin Speech Synthesis Database. Sentence 050032, from 4.155 sec to 4.190 sec.

The resonance frequency is now 0.754 kHz, and the corresponding Euler tube length is $352/(4\times0.753) \approx 117$ mm, which is about right.

As mentioned in Section 2.2.5, because the voluntary force to start a plosive and the Bernoulli force are in opposite directions, the situation is unstable. Sometimes, a trilling of the tongue against the palate occurs. Multiple excitations of Euler resonance waves during the starting of a single plosive are observed. Figure 2.22 shows the observation of three consecutive Euler resonance waves in a single plosive of [kʌ].

2.3.2   Fricatives

Fricatives are also universal in all languages of the world. As an example, the positions of the oral cavity during the enunciation of fricatives [s] in "see" and "soon" is shown.

The source of the sizzling sound can be identified by doing a five-second experiment by yourself, as shown in Fig. 2.23. First, pronounce a voiceless fricative with devoiced [i] in "see", as in Fig. 2.23(A). The sound comes from the space between the tip of tongue and the teeth. Next, pronounce a voiceless fricative with devoiced [u] in "soon", as in Fig. 2.23(B). The sound comes from the space between the tip of tongue and the protruding lips. The difference of the vibrational frequencies can be perceived. During the enunciation of either sizzling consonants, open your mouth but do not change anything else. The consonant immediately stops. There is no sound coming from the throat.

The experimentally observed amplitude spectrum of the signal further clarifies the source. Figure 2.24(A) shows the spectrum of [si]. It is generated by cutting a 5 msec portion of the signal, take the Fourier transform, then smooth it over a 0.1 kHz interval. Four such examples are displayed. The peak of the spectrum is at 5.5 kHz. Figure 2.24(A) shows the spectrum

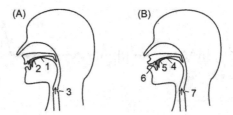

Fig. 2.23. Oral cavity geometry for fricative [s]. (A) In "see", the space between the hard palate and teeth makes an Euler transient resonator. (B) In "soon", the length of the resonator is increased.

of [si]. As shown, the peak shifts to 4.4 kHz. Additional features around 10 kHz are also observed.

The observed spectrum can be understood in terms of Euler tube resonators, see Fig. 2.23. For [s] in "see", the constriction at point 1 of Fig. 2.23(A) is the source of turbulence, which can be considered as a random emission of pulses. Each pulse excites a resonance wave in the Euler transient resonator 2. The peak frequency 5.5 kHz is associated with a length 16 mm, which is about right as the distance between the constriction 1 and the teeth 3. For [s] in "soon", Fig. 2.23(B), the resonator is longer. The peak frequency, 4.4 kHz, is associated with a length 20 mm, which is about right as the sum of cavities 5 and 6. The features around 10 kHz are associated with shorter Euler transient resonators, probably 6 in Fig. 2.23(B).

A significant difference of the fricative from vowels is the phase spectrum, as shown in Fig. 2.25. It is a universal observation that the phase spectrum

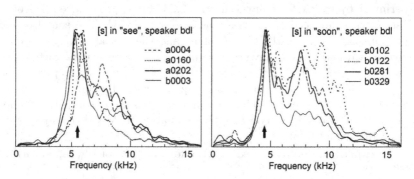

Fig. 2.24. Amplitude spectra of fricative [s]. (A) In "see", the peak frequency is 5.5 kHz. (B) In "soon", the peak frequency is 4.5 kHz. Additional peaks around 10 kHz are observed, originating from shorter resonance tubes.

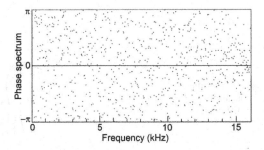

Fig. 2.25. Phase spectrum of fricative [s]. A universal observation of the fricatives is that the phase spectrum is completely random, reflecting the turbulent nature of the excitation arisen at the constriction of the tube resonator.

of fricatives is completely random, reflecting the turbulent nature of the excitation arisen at the constriction of the tube resonator.

2.3.3 Vowels

While plosives and fricatives are produced solely in the oral cavity, vowels and voiced consonants are produced jointly by the vocal folds and vocal tract. In 1829, Robert Willis published the first detailed study of the mechanism of vowel production [99]. As a capable mechanical engineer, Willis built a number of mechanical models to demonstrate the production of vowel sounds, a few of them are illustrated in Fig. 1 in the Preface. At the incoming end of the demonstration device, he used a reed to mimic the vocal folds to generate periodic air puffs at a given pitch frequency. A tube connected to the reed mimics the vocal tract. Willis applied Euler's theory of transient resonators for explanation: each time a pulsation is emitted by the reed, it will resonate back and forth in the tube to form a decaying wave. The wavelength of the decaying wave is four times the length of the tube. Willis showed that by changing the length of the tube, sound of different colors can be produced to mimic different vowels. With a fixed tube length, while the pitch frequency is changed by the structure of the reed, the timbre of the vowel is unchanged. In other words, he showed that the character of the vowel only depends on the resonance tube, not the reed. The single resonance frequency model of Willis is definitely an oversimplification for most vowels. However, for a few vowels, there is a dominant format, and the frequency of that dominant formant can be explained by simplifying the vocal tract as a tube with uniform cross section, such as [u] and [ɑ]. Note that this intuitively inspiring model is for conceptual understanding only. In reality, the structure of the vocal tract is much more complicated, and a rich array of formants can be produced. See Chapter 4 for details.

Fig. 2.26.   Oral cavity while enunciating [u].   The lips are rounded, and the tongue is lowered.   The entire length of the upper vocal tract forms an Euler transient resonator.   For male speakers, the length is about 190 mm.   The resonance frequency is about 460 Hz.

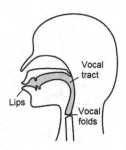

First, we study the production of vowel [u], as shown in Fig. 2.26. During the enunciation of vowel [u], the lips are rounded to form a small orifice of diameter about 10 mm, and the tongue is lowered. The entire length of vocal tract, from the vocal folds to the lips, forms an Euler transient resonator, see Fig. 2.26. For male speakers, the average length of the vocal tract is 169 mm. With protruded lips, the total length is about 190 mm. According Euler's formula, Eq. 1.34, the resonance frequency is

$$f = \frac{c}{4L} = \frac{352000}{4 \times 190} \approx 460 \, \text{Hz}. \qquad (2.24)$$

Figure 2.27(A) shows the waveform of a vowel [u] pronounced by a male speaker with pitch frequency about 100 Hz, which is typical. As shown, each pitch period starts with a strong peak, then gradually decays. The waveform in each pitch period oscillates four to five times. Figure 2.27(B) is the amplitude spectrum of a single pitch period. As shown, except for the peak at about 100 Hz which is the fundamental-frequency component, the main feature of the spectrum is a strong peak at about 460 Hz.

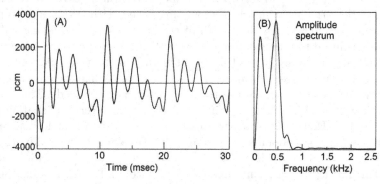

Fig. 2.27. Waveform and amplitude spectrum of [u]. (A) Waveform of vowel [u]. (B) Amplitude spectrum of [u], showing a main formant peak at 460 Hz.

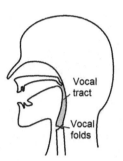

Fig. 2.28.   Oral cavity while enunciating [ɑ]. The oral cavity is widely open. The narrow tube between the glottis and the soft palate becomes an Euler resonator. The typical length of that section is about 110 mm, corresponding to a resonance frequency of about 800 Hz.

Next, we study the production of vowel [ɑ], see Fig. 2.28. While enunciating vowel [ɑ], the mouth is widely open, and the back end of the tongue is close to the soft palate. The mouth is effectively a part of the open air. The pharynx is essentially the resonance tube. The typical length is 110 mm. The expected resonance frequency is 800 Hz. Figure 2.29(A) shows an observed waveform. In each pitch period, there are about 8 decaying sinusoidal cycles. Figure 2.29(B) shows that the vocal-tract resonance frequency is nearly 800 Hz, in line with the expectation. The actual geometry of the vocal tract is more complicated, which gives rise to the second formant of the vowel [ɑ] near 1400 Hz and other features.

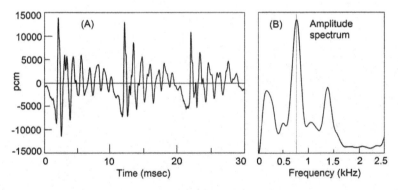

Fig. 2.29. Waveform and amplitude spectrum of [ɑ]. (A) Waveform of vowel [ɑ]. (B) Amplitude spectrum of [ɑ], showing a main formant peak at 800 Hz.

Chapter 3

Experimental Facts

In this chapter, we present basic experimental facts of human voice, collected by various instruments. Based on those facts, in Chapter 4, the physics of human voice production is expounded, which becomes the foundation of the mathematical representations of human voice in Part II.

3.1 Microphones and Voice Signals

The most abundant experimental data of human voice are collected by microphones. In this section, we present the working principles of microphones and observations from voice signals collected by microphones.

3.1.1 Types and working principles of microphones

To understand the nature of microphone signals and to evaluate its quality, the modern microphones and their working principles are presented. Early types of microphone, such as the carbon microphone, are not presented here because of the signal quality is low. The microphones used in modern human-voice recordings are of two categories, the condenser microphone and the dynamic microphone. There are two types of condenser microphones. The membrane type, requiring a *phantom power*, is used in high-quality recordings. The electret type is used in portable electronics.

The condenser microphone, invented in the early 20th century, is currently the standard microphone for high-quality voice recordings [36, 51, 100]. Photographs of a typical condenser microphone is shown in Fig. 3.1. Figure 3.2 shows its working principle and frequency response curve. Figure 3.1(A) is an internal view after the cover is removed. The sensor of the microphone is a thin membrane plated with gold, typically 5 μm thick, extended on a frame with a tension of around 2000 nN/m against a perforated backplate with an air-gap of typically 25 μm, see Fig. 3.2(A). A capacitor of about 50 pF is formed. A DC voltage source, called phantom power and typically 48 V, is applied to polarize the capacitor. Utilizing the available DC power, a built-in preamplifier is mounted inside the cover, as shown in Fig. 3.1(A) and Fig. 3.2(A). The output impedance is converted

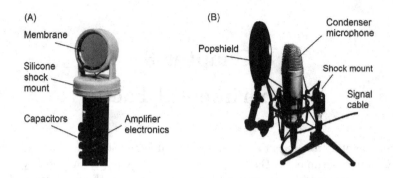

Fig. 3.1. Condenser microphone. Photographs of a condenser microphone Røde NT1-A, taken in the author's recording studio. (A). Internal view without the cover. The diameter of the membrane is 25 mm. A silicone shock mount and a preamplifier are inside the cover. (B). A complete setup with popshield and external shock mount.

from as high as 10 MΩ to as low as 100 Ω, to make it immune to external electrical noise. The microphone is protected against mechanical vibration by an internal shock mount made of silicone, as well as an external shock mount, as shown in Fig. 3.1. The sensitivity of condenser microphones is typically 25 to 50 mV/Pa. The frequency response, as shown in Fig. 3.2(B), is within a few dB from 20 Hz to 20 kHz.

The condenser microphone is extremely sensitive to any motion of air.

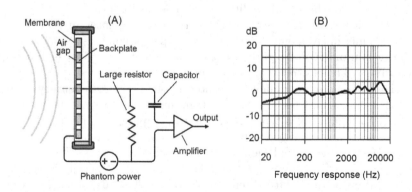

Fig. 3.2.   Working principle and frequency response of condenser microphone. (A). Working principle. A 5 μm thick membrane is extended on a frame against a perforated backplate with a 25 μm air gap to form a capacitor. The membrane responds to acoustic pressure to generate a voltage, which is then amplified by the built-in preamplifier. (B). Frequency response. Source: www.rodemic.com.

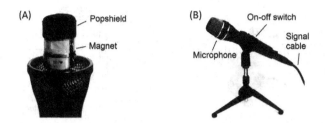

Fig. 3.3.   Dynamic microphone. Photographs of a Sennheiser MD 431 dynamic microphone, taken in the author's recording studio. (A). An internal view after the cover is removed. (B). The setup including a mini-tripod.

While pronouncing stop consonants such as [p], [t], and [k], the air jet can generate a wild gyration in the output signal. For human voice recording, a *popshield* is installed in front of the microphone, see Fig. 3.1(B).

Another important type of microphone is dynamic microphone. It is more rugged than the condenser microphones and does not require a phantom power. Therefore, it is used more often than the condenser microphone. Figure 3.3 shows photographs of a typical dynamic microphone, and Fig. 3.4 shows it working principle and frequency response. Figure 3.3(A) shows an internal view after the cover is removed. Although it is much less sensitive to the motion of air than the condenser microphone, an internal foam-plastic pop shield is built in, see Fig. 3.4(A). The dynamic microphone often includes a radio-frequency emitter to make it wireless, and it can then be held by hand during recording. As shown in Fig. 3.4(A), the basic structure of the dynamic microphone is identical to a dynamic loudspeaker. When the coil moves in a magnetic field, an electromotive force is generated. As the

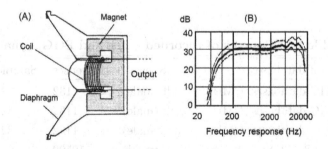

Fig. 3.4. Working principle and frequency response of dynamic microphone. (A). Working principle. When the coil moves in a magnetic field, an electromotive force is generated. (B). Frequency response. Source: Brochure of MD 431 microphone.

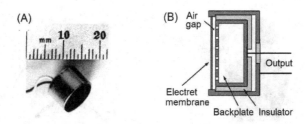

Fig. 3.5.   Electret microphone. (A). Photographs of a MCE-100 microphone, taken in the author's recording studio. (B). Internal structure of an electret microphone.

condenser microphone responds to air particle displacement, the dynamic microphone responds to air particle velocity. As shown in Fig. 3.4(B), at frequencies lower than 100 Hz and higher than 15 kHz, the response of dynamic microphone becomes lower.

For portable electronic devices, including laptop computers and smartphones, the most common type of microphone is *electret microphone*. A photograph of a MCE-100 electret microphone is shown in Fig. 3.5(A). The working principle is identical to the studio condenser microphone. Instead of using a phantom power to polarize the capacitor, a thin membrane of permanently polarized material, the *electret* (a word coined by combining *electr*ic and magn*et*), to form an electric field inside the air gap between the membrane and the baseplate, see Fig. 3.5(B).

3.1.2   Source of voice signals

The voice samples used in this book are listed in Table 3.1, recorded with high-quality microphones under well-controlled conditions by professional speakers, with simultaneous electroglottograph signals.

Table 3.1: Sources of recorded voice and EGG signals

Source of database	Speaker	Sentences	Sample rate
ARCTIC databases, CMU	bdl (male)	1132	32 kHz
ARCTIC databases, CMU	jmk (male)	1132	32 kHz
ARCTIC databases, CMU	slt (female)	1132	32 kHz
King-TTS-003, Speechocean	female	19509	44.1 kHz
King-TTS-012, Speechocean	male	15000	44.1 kHz

3.1.3 Vowels

An essential feature of human voice is that the pitch frequency, defined as the inverse of the pitch period, is constantly varying. The production process of vowels can be understood by analyzing experimental facts of vowels under varying pitch frequencies. The examples shown here are sections of vowels near the end of a breathing unit. In each section, the pitch value starts in the high or middle part of the speaker's pitch range, then gradually lowers. There are thousands of such events in the databases of Table 3.1.

The exposition of information in each of the figures is as follows: In part (A), the top chart is the raw waveform in PCM, the middle chart is the time derivative of PCM, and the lower chart is short-time power averaged over every two msec displayed in dB scale. The last four periods in the speech signal are segmented and ends-matched; then a Fourier analysis is executed on each pitch period, shown in (B). The peaks in the amplitude spectra, that is, the formant frequencies, are marked on each peak.

Vowel [ɑ]. Figure 3.6 shows the experimental data of vowel [ɑ]. At the beginning of the time interval, the pitch frequency is 184 Hz, in the middle-high range of the male voice. The pitch frequency gradually decreases to 118 Hz. Inside each pitch period, the short-time averaged power gradually decreases as well. The dB chart clearly shows an exponential decay within each pitch period with a fairly consistent rate of −5 dB per msec. The amplitude spectra of the last four pitch periods, Fig. 3.6(B), are almost identical despite huge pitch period differences. The waveforms of the earlier pitch periods strikingly resemble the starting part of the waveform of later periods. If an early waveform is allowed to evolve freely, each one will display a full decaying wave, resembling the last one.

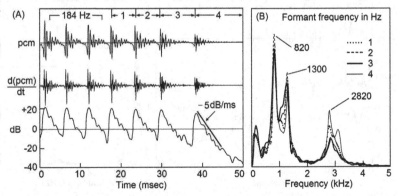

Fig. 3.6. Vowel [ɑ]. Source: Sentence 050007 of King-TTS-012, 2.23 sec to 2.28 sec. (A). Waveform and short-time power. (B). Amplitude spectra of four pitch periods.

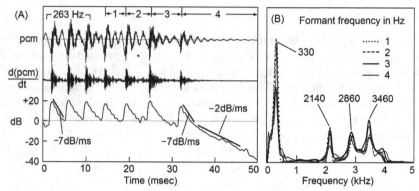

Fig. 3.7. Vowel [i]. Source: Sentence 004419 of King-TTS-012, 1.938 sec to 1.988 sec. (A). Waveform and short-time power. (B). Amplitude spectra of four pitch periods.

Vowel [i]. Figure 3.7(A) shows the experimental data of vowel [i]. At the beginning, the pitch is 263 Hz, in the high end of the speaker's pitch range. The decay of the short-time power, in dB scale, has two stages. At the beginning of a pitch period, the short-time power decays −7 dB per msec, which is the decay rate of the 2-3 kHz formants. Later, the decay rate becomes −2 dB per msec, that of the 330 Hz formant.

Vowel [u]. As shown in Fig. 3.8, the experimental data of vowel [u] displays features similar to vowels [ɑ] and [i]. The starting pitch, 253 Hz, is in the high end of the speaker's pitch range. The dominant features in the amplitude spectrum are two formants in 400 Hz and 790 Hz . The short-time power decays at a constant rate of -2dB/msec.

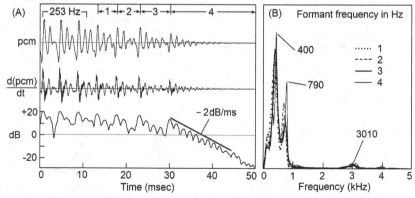

Fig. 3.8. Vowel [u]. Source: Sentence 005044 of King-TTS-012, 1.06 sec to 1.11 sec. (A). Waveform and short-time power. (B). Amplitude spectra of four pitch periods.

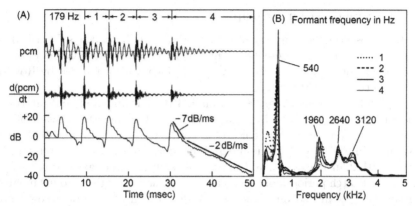

Fig. 3.9. Vowel [e]. Source: Sentence 050053 of King-TTS-012, 2.535 sec to 2.585 sec. (A). Waveform and short-time power. (B). Amplitude spectra of four pitch periods.

Vowel [e]. As shown in Fig. 3.9, the experimental data of vowel [e] displays features similar to vowel [i]. There are three high-frequency formants which decay at −7 dB/ms within each pitch period, and the 540-Hz formant decays at −2 dB/ms within each pitch period. The corresponding amplitude spectrum shows a sharp formant peak at 540 Hz.

Vowel [o]. As shown in Fig. 3.10, the experimentally observed features are similar to those of vowel [u]. The starting pitch, 161 Hz, is in the middle-high portion of the speaker's pitch range. The amplitude spectrum shows a sharp formant at 620 Hz. The short-time power shows a consistent −2 dB/ms decay rate, apparently related to that formant.

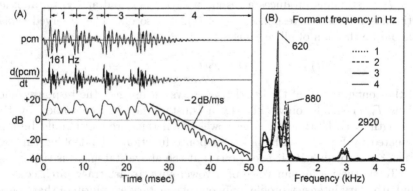

Fig. 3.10. Vowel [o]. Source: Sentence 051022 of King-TTS-012, 1.827 sec to 1.877 sec. (A). Waveform and short-time power. (B). Amplitude spectra of four pitch periods.

3.1.4   Superposition principle

In this subsection, a summary of the experimental observations described in the previous section is presented. The experimental facts can be understood in the light of the *superposition principle*, which was formulated by Robert W. Scripture in 1930 [84].

By looking through Figs. 3.6 through 3.10, it is apparent that although the pitch period changes by a factor of two, the underlying elementary waveforms representing the color of a vowel are essentially identical. For example, in Fig. 3.7(A), the number of first-formant cycles in a pitch period starts with 2, increases to 3 and then 4. Nevertheless, the two feature groups in the first pitch period look identical to the first two feature groups in later pitch periods, and so on. The amplitude spectra of the same vowel are essentially identical while the pitch varies by more than an octave.

In Figs. 3.6 through 3.10, it is clear that for each pitch period, the signal starts with a strong impulse, then decays. The high-frequency components between 2 kHz to 4 kHz decay faster, at a rate of -7 dB/ms. The medium frequency components, around 1 kHz, decay more slowly, at -5 dB/ms. The low-frequency components, between 0.3 kHz and 0.5 kHz, decays even more slowly, at -2 dB/ms. Again, if those formant oscillations were allowed to evolve freely, each one would continue until it disappears.

To summarize, for each pitch period, the signal starts with a strong impulse, then decays. For each vowel, the underlying elementary waveforms are essentially identical. The entire signal can be constructed by the *superposition principle*, which is a property of linear systems.

Because the wave equation is linear, the superposition principle should be valid. It states that the response caused by two or more stimuli is the sum of the responses caused by each stimulus individually. Assuming at $t = t_1$, a stimulus produces a response $F_1(t - t_1)$, and at $t = t_2$, another stimulus produces a response $F_2(t - t_2)$, and so on. The observed signal should be the sum of all responses,

$$S(t) = F_1(t - t_1) + F_2(t - t_2) + F_3(t - t_3) \ldots . \tag{3.1}$$

If the configuration of the vocal tract stays unchanged, the response functions $F_n(t)$ should only differ by a constant factor, determined by the strength of each stimulus. The vowel signal thus produced could be represented by Eq. 3.1 with similar response functions. By looking through Figs. 3.6 through 3.10, it is obvious that the above statement is true.

To synthesize human voice of a vowel with an arbitrary pitch contour and arbitrary intensity profile, only one elementary waveform of that vowel, allowed to decay to the end, is needed. Figure 3.11 shows a simple case with identical intensity and equal time interval between the starting moments.

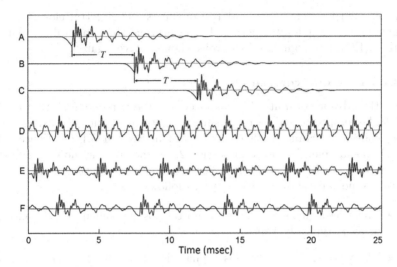

Fig. 3.11. Superposition principle. (A) through (C), three elementary waves of vowel [e], with starting time separated by a pitch period T. Waveforms extracted from recorded voice, as displayed in Fig. 3.9. (D) through (F), synthesized voice using the superposition principle with different pitch values. (D), C4, 261.63 Hz, $T = 3.82$ msec. (E), G3, 196.0 Hz, $T = 5.10$ msec. (F), C3, 130.81 Hz, $T = 7.64$ msec.

The elementary wave, here for a vowel [e], is extracted from a recorded voice, as displayed in Fig. 3.9. (A) through (C), three elementary waves of vowel [e], each separated by a pitch period T. (D) through (F) show synthesized voice using the superposition principle with pitch values varying over one octave. The synthesized voice is smooth, but sounds robotic because of the constant pitch. Using Eq. 3.1, with varying pitch periods and intensities, more human-like voice can be synthesized, see Chapter 8.

The superposition principle presented here is phenomenological: it is solely based on observations on recorded voice signals. For a better understanding of the underlying physics, especially the nature of the stimulus that starts the elementary wave of a vowel, other experimental facts need to be studied. These are presented in the following sections.

3.2 Electroglottograph and Voice Data

In Section 2.2.4, the working principle and observations of the electroglottograph (EGG) are presented. As shown, the peaks in the derivative of the EGG signal point to the closing moment and opening moment of the glottis. As a result of the Bernoulli force, the d(EGG)/dt signal at glottal closings is much sharper than that at glottal openings. The glottal closing is the defining moment of a pitch period. In this Section, the temporal correlation of the d(EGG)/dt signal and the voice signal is investigated.

3.2.1 Temporal correlation

Since the advance of unit-selection speech synthesis research started in the 1980s, large speech databases of simultaneous voice and EGG data have been collected. Some of them are shown in Section 3.1.2. Figure 3.12 shows an example. As shown in Section 2.2.4, the opening moment and the closing moment divide each pitch period into a *closed phase* and an *open phase*. Some general observations are as follows:

- The voice signal starts after the glottal closing moment with a constant delay τ, typically 1 msec.

- The voice signal in the closed phase is much stronger than the voice signal in the open phase.

- The effect of opening moment on the voice signal is minor and variable. Sometimes it causes mild noise signals.

The observed facts can be explained by looking at Fig. 3.12. The linear distance from the glottis to the microphone, marked on Fig. 3.12 as a gray curve L, is about 350 mm. The time for an acoustic pulse to travel from

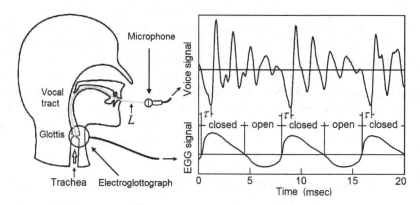

Fig. 3.12. Microphone signals and electroglottograph signals. At a glottal closing moment, an acoustic pulse is generated. The time for an acoustic pulse to travel from the glottis to the microphone through the curve L is about 1 msec. The voice signal in the closed phase is strong, and the voice signal in the open phase is weak. Source: ARCTIC databases, speaker bdl, sentence b0274, time 1.454 sec to 1 476 sec.

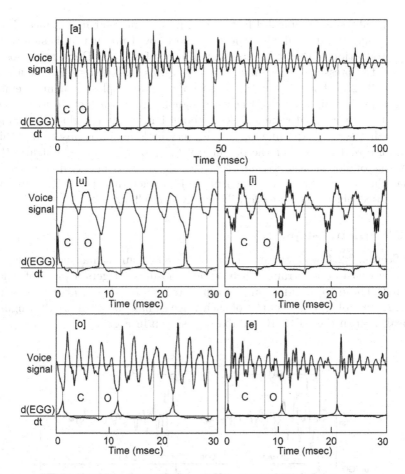

Fig. 3.13. Combined voice and EGG signals. Source: ARCTIC databases, speaker bdl. (A), vowel [a], sentence a0016; time 1.24 sec to 1.34 sec. (B), vowel [u]. (C), vowel [i]. (D), vowel [o]. (E), vowel [e]. In all cases, the time delay from the glottal closing moment to the starting of the voice signal is about 1 msec, which is associated with the propagation time of an acoustic signal from the glottis to the microphone.

the glottis to the microphone is about 1 msec. Therefore, the *glottal closure* is the origin of the stimulus. In other words, at the time the glottis closes, an impulse emerges from the glottis and starts the elementary wave of the vowel of each pitch period. That impulse takes about 1 msec to travel along the gray curve to reach the microphone. Figure 3.13 shows the correlation between voice signal and EGG signal for six vowels. The above facts are universally observed.

The above experemental facts are counterintuitive at first glance. Naturally, one would think that the process of vowel production is similar to the production of plosive [k], presented in Section 2.3.1: During the closed phase, the pressure in the trachea builds up, as marked by a hollow arrow in Fig. 3.12. It forces the vocal folds to open in an explosive manner, emitting a puff of air to excite the voice signal. If such a picture were true, then the voice signal would peak at about 1 msec after a glottal *opening*. Nevertheless, the experimental fact is, very little action is observed within a millisecond or so after the opening of the glottis. On the contrary, the opening of glottis often *weakens* the voice signal. It is a general observation that the voice signal in the open phase is weaker than the voice signal in the closed phase. This puzzle will be solved in Chapter 4.

3.2.2   Glottal stops

In Section 2.2.4, we have shown isolated EGG closure signals not connected with a quasi-period train. Each of such closures triggers a piece of voice signal. These voice signals are known as the *glottal stop* in phonetics, with a IPA symbol [ʔ]. An example of such voice signals, by a female US English speaker slt in the ARCTIC database, is shown in Fig. 3.14.

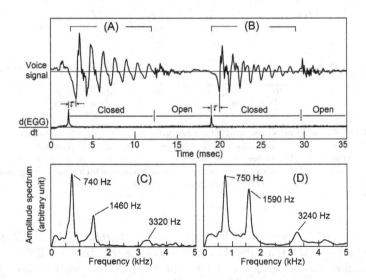

Fig. 3.14.   Glottal stops. Source: ARCTIC databases, speaker slt, sentence a0312, time 1.521 sec to 1.557 sec. The sections of waveform (A) and (B) are Fourier analyzed to generate amplitude spectra (C) and (D). The formant structure is essentially identical to that of the typical formant structure of vowel [ɑ].

In Fig. 3.14, two consecutive glottal stops, with a time elapse if 17 msec, are shown. From a perception point of view, the two events are perceived as a single glottal stop, similar to the multiple velar plosive shown in Section 2.3.1. The correlation of the glottal closure detected by EGG signal and the starting of the glottal stop signal is identical to the production of vowels, see Section 3.1.3: a delay of about 1 msec originated from the propagation time of the line L in Fig. 3.12. In the closed phase, clean decaying signals are observed. In the opening moment, certain noise is generated, also similar to the phenomenon observed in vowels.

The segments of waveforms extracted from the signal, marked as (A) and (B), are Fourier-analyzed, and the amplitude spectra are shown as (C) and (D), respectively. As shown, the formant peaks match well with those of vowel [ɑ]. Clearly, although those events are characterized as stop consonants, the spectral nature is identical to the vowel [ɑ], see Fig. 3.6. If those events are allowed to repeat with a time interval of for example 5 msec, a sustained vowel [ɑ] is produced.

3.3   In-Vivo Pressure Measurements

In the 1960s, semiconductor miniature pressure transducers for medical research were developed. Those pressure transducers, mounted on a flexible catheter of 2 mm in diameter, originally designed for inserting into the blood vessels, with a frequency response up to 20 kHz, are ideal to probe into the larynx to study the mechanisms of human voice production. A number of reports of subglottal pressure and supraglottal pressure measurements during phonation were published[1] showing informative temporal correlations with simultaneously acquired EGG signals [21, 61, 82].

Figure 3.15 shows a typical configuration of experimental setup. The pressure sensors are inserted into the larynx through the nasal cavity. One pressure sensor is placed beneath the vocal folds, and another one is placed above the vocal folds. Therefore, the variation of subglottal pressure and the supraglottal pressure can be measured simultaneously with a time resolution of better than 0.1 msec while vocalizing. The voice signals through a microphone and the EGG signals can be acquired simultaneously. To ensure that the measurements are synchronized, a multiple-channel data recording device is utilized.

The most salient and persistent feature of the observed subglottal pressure and supraglottal pressure is the temporal coincidence of maximum sub-

---

[1]In an article by Yasuo Kioke on page 181 of *Vocal Fold Physiology* [87], several reports on sub- and supraglottal pressure variation measurements were cited. Nevertheless, no simultaneous EGG signals were acquired and compared.

glottal pressure and minimum supraglottal pressure with the glottal closing moment identified by EGG signals. Typical results are shown in Fig. 3.16. As shown by lines marked A, A', and A", the maximum subglottal pressure and the minimum supraglottal pressure occur simultaneously at a glottal closure instant. Furthermore, the subglottal pressure at the glottal closure moment is the *absolute maximum pressure* in the entire pitch period, and the supraglottal pressure at the glottal closing moment is the *absolute minimum pressure* in the entire pitch period.

On the other hand, around each glottal opening moment as detected by EGG signals, very weak variations are noticed in both subglottal pressure and supraglottal pressure. As shown in Fig. 3.16, before the glottal opening moments, as marked by B and B', the subglottal pressure is at a relatively low value, that is, in the lower range of the pressure scale. After the glottal opening, immediately to the right side of the lines marked as B and B', the subglottal pressure does not change noticeably.

Another remarkable observation is that at each glottal closing moment, the absolute value of the subglottal pressure increase at the glottal closing moment roughly equals the absolute value of the supraglottal pressure decrease. And the rate of pressure change during the glottal closing moment

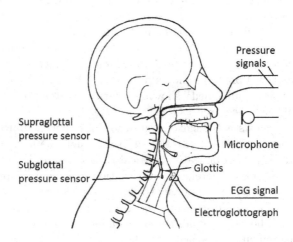

Fig. 3.15. In-vivo pressure measurement apparatus. Two miniature semiconductor pressure sensors, one supraglottal and one subglottal, are placed using a flexible catheter through the nasal cavity. Pressure reading with time resolution down to 0.1 msec can be acquired simultaneously with voice signal and EGG signal. Using a multiple-channel data recording device, those measurements are well synchronized. After Schutte and Miller [82].

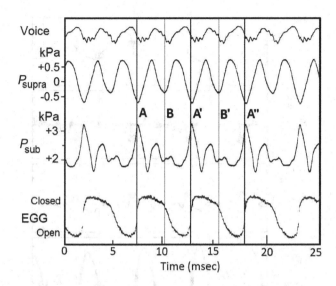

Fig. 3.16.   Time-sequence of glottal pressures. The maximum of subglottal pressure and the minimum of supraglottal pressure occur at the glottal closing moment, A, A' and A". Near the moments of glottal opening, marked by B and B', the subglottal pressure is in the lower range of the pressure scale, and does not change much near a glottal opening. The absolute magnitude of the rise of subglottal pressure, about 1 kPa, is approximately equal to that of the drop of supraglottal pressure. After Miller and Schutte [61].

is the greatest within the entire pitch period.

The above observations are confirmed by the measurements of another research group, shown in Fig. 3.17. In that Figure, the glottal closing moments are marked by an asterisk "*", and the glottal opening moments are marked by a circle "o". As shown, at the glottal closing moments, the subglottal pressure reaches to an absolute maximum in each pitch period, and the supraglottal pressure reaches to an absolute minimum. At the glottal opening moments, the pressure remains in a moderate value with very little variation around the moment of glottal opening.

The results of a follow-up experiment on five different vowels are shown in Fig. 3.18. For all vowels, the maximum subglottal pressure occurs at glottal closure instants, and the waveforms of subglottal pressure are similar for all vowels. Nevertheless, *the waveforms of the supraglottal pressure show noticeable difference for different vowels.* That fact can be understood from the nature of the pressure measurement: After the first valley of supraglottal pressure at the glottal closing moment, the timing of the next valley of the supraglottal pressure should be the moment when the acoustic wavefront

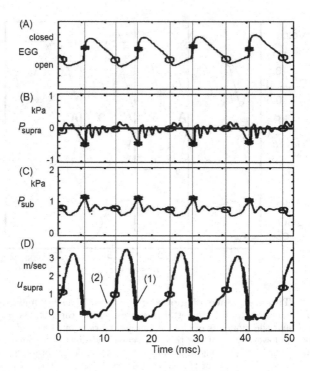

Fig. 3.17.   Pressure measurements for vowel [a]. The glottal closing moments are marked by an asterisk *, and the glottal opening moments are marked by a circle o. Here, (A) is the EGG signal, (B) is the supraglottal pressure, (C) is the subglottal pressure, and (D) is the speech of airflow immediately above the glottis. As shown, at the glottal closing moments, the subglottal pressure reaches to an absolute maximum in each pitch period, and the supraglottal pressure reaches to an absolute minimum. At the glottal opening moments, the pressure remains in a moderate value with very little variation After Cranen and Bovis [21].

reflecting from the vocal tract arrives. For vowel [a], the time is shorter, and for vowel [u], the time is longer, see Section 2.3.3. For the waveforms of the subglottal pressure, the time of pressure peak following the first pressure peak at the glottal closing moment is determined by the time the acoustic wave reflecting from the lower end of the trachea arrives. Because the length of the trachea does not change with the identity of the vowel, the waveforms of the subglottal pressure do not depend on the vowel, as expected.

For more than one century, there has been a popular opinion that in each pitch period, a vowel sound is excited by a puff of air at the opening moment of the glottis [23]. When the glottis is closed, the pressure from the lungs and trachea builds up. At a certain point of time, the subglottal air

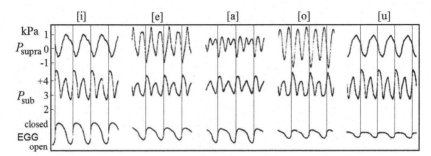

Fig. 3.18. **Pressure variation of several vowels.** After the initial surge, the subsequent evolution of subglottal pressure for all vowels are similar. However, after the initial drop, the supraglottal pressure for different vowels diverge significantly, indicating an acoustic effect depending on the instantaneous geometry of the vocal tract. After Schutte and Miller [82].

pressure becomes so high that it forces the vocal folds to separate rapidly, releasing an explosive puff of air. And that puff of air at the glottal opening moment is the source of vowel signals. According to such point of view, immediately before a glottal opening, the subglottal pressure should be the highest. Immediately after a glottal opening, the subglottal pressure should drop quickly to a minimum. None of those phenomena are observed experimentally. This is because that popular opinion is based on an intuition in *fluid mechanics*. In reality, the processes in the trachea and vocal tract near the vocal folds are governed by the law of *acoustics*, or more precisely, by the *wave equation*, rather than by fluid mechanics. In Chapter 4, by resolving the wave equation with proper boundary conditions and proper initial conditions in the vocal tract and in the trachea, a quantitative explanation of the observed pressure variations is provided.

Chapter 4

The Physics of Voice Production

In Chapter 3, the basic experimental facts of human voice are presented. In this Chapter, a theory is formulated to explain the experimental observations as completely as possible. The focus is vowels and voiced consonants, especially to explain voice waveforms with simultaneous EGG signals, videokymography information, and *in vivo* pressure measurements. The production of plosives and fricatives is relatively straightforward, which has been outlined in Section 2.3 in Chapter 2.

4.1 A Brief Summary of Experimental Facts

Following is a brief summary of the experimental facts of vowels:

1. The vowel signals satisfy the superposition principle. A segment of vowel signal can be expressed as a linear superposition of individual elementary waves F_n started at time t_n,

$$S(t) = F_1(t - t_1) + F_2(t - t_2) + F_3(t - t_3) + ... \qquad (4.1)$$

2. Each elementary wave starts at a glottal closing moment.

3. The glottal closing is abrupt and forceful.

4. The voice signal is strong in the closed phase.

5. The glottal opening is slow, weak, and sometimes noisy.

6. The voice signal is weak in the open phase.

7. The higher-frequency components of the voice signal (1 kHz to 4 kHz) concentrate at the beginning of the closed phase and decay faster.

8. The lower-frequency components of the voice signal (<1 kHz) decay more slowly and dominate the later portion of the elementary wave.

9. High closed quotient corresponds to high voice intensity.

10. Low closed quotient corresponds to low voice intensity.

11. The subglottal pressure rises to a maximum at glottal closings.

12. The supraglottal pressure drops to a minimum at glottal closings.

Obviously, the key to understand the production of vowels is the acoustic process in the vocal tract after a glottal closing. We present a conceptual picture first, then present a rigorous mathematical treatment, followed by explanations of various experimental facts.

4.2   The Concept of Timbrons

The experimental facts can be explained by the following theory:

1. Immediately before a glottal closing, there is a steady and uniform airflow in the vocal tract with a typical velocity of 0.5 m/s to 5 m/s. The airflow varies slowly, the resulting acoustic excitation is weak.

2. A glottal closing abruptly terminates the airflow from the trachea into the vocal tract. Because of inertia, air in the vocal tract continues to flow with the steady velocity.

3. A zero-particle-velocity wavefront is created at the glottis, propagating at the speed of sound into the vocal tract, converting the aerodynamic kinetic energy of the airflow into acoustic energy.

4. The above elementary acoustic wave resonates in the vocal tract and radiates. Its waveform is determined by the geometrical configuration of the vocal tract, thus representing the instantaneous timbre.

5. The glottal closing does not supply energy to the voice. It *triggers* the conversion of the kinetic energy of the steady airflow immediately before a glottal closing in the vocal tract into acoustic energy.

6. A glottal opening connects the lungs to the resonance cavity of the vocal tract, often accelerating the decay of the acoustic wave.

7. If a glottal closing takes place before the previous elementary acoustic waves disappear, the acoustic wave triggered by the new glottal closing superposes on the tails of the previous elementary acoustic waves.

The waveform of the elementary acoustic wave triggered by a glottal closing is determined by the instantaneous geometrical configuration of the

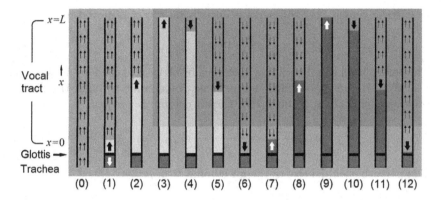

Fig. 4.1. Acoustic waves triggered by a glottal closure. (0), the glottis is open, there is a uniform and steady airflow in the trachea and the vocal tract. (1) through (3), a glottal closing generates a zero-particle-velocity d'Alembert wavefront propagating towards the lips. (4), the d'Alembert wavefront reflects at the lips with a reversed air-particle velocity. (4) through (6), the reversed d'Alembert wavefront propagates towards the glottis. (7), the d'Alembert wavefront reflects at the glottis to generate a d'Alembert wavefront of compressed air. (7) through (9), the d'Alembert wavefront of compressed air propagates towards the lips. (10), the d'Alembert wavefront reflects at the lips to generate a d'Alembert wavefront with particle velocity pointing to the open air. (10) through (12), the d'Alembert wavefront propagates towards the glottis to recover the initial state similar to (1).

vocal tract thus it represents the timbre at that moment. The above elementary wave is reasonably to be termed as a *timbron*. Timbrons are elementary building blocks of voice, in a similar sense as atoms are elementary building blocks of matter. For convenience, the theory presented here is termed *timbron theory*, in a similar sense to that of the atomic theory of matter.

To illustrate the above concept step by step, a single-tube model of the vocal tract shown in Fig. 4.1 is analyzed. The medium gray background indicates the unperturbed air density ρ_0. The light gray area represents rarefied air particles. The dark gray area represents compressed air particles. The small arrows indicate the particle velocity, initially u_0. The large arrow indicates the velocity of sound. Steps (0) through (12) are marked.

Let us trace the process step by step. Before a glottal closing, step (0), a steady airflow with velocity u_0 runs from the trachea through the glottis and the vocal tract into open air. The density of air over the entire length is uniform. Immediately after a glottal closing, (1), the velocity of air at the glottis is abruptly forced to zero. Because the air particles keep moving up with velocity u_0, a zero-particle-velocity wavefront is created and propagates into the vocal tract. At time t, the wavefront is at $x = ct$. Due to particle velocity u_0, the particle initially at $x = ct$ is moved to

$$x' = (c + u_0)t. \tag{4.2}$$

The air column left behind by the wavefront is expanded by a factor of $1 + u_0/c$. Because no additional air mass could come through the glottis, conservation of mass requires that the air column is *rarefied*,

$$\rho_0 \to \frac{\rho_0 \, x}{x'} = \frac{ct \, \rho_0}{(c + u_0)t} \approx \rho_0 - \rho_0 \frac{u_0}{c}. \tag{4.3}$$

The perturbation density of the air column is *negative*,

$$\rho = -\rho_0 \frac{u_0}{c}. \tag{4.4}$$

The relation between perturbation pressure and perturbation density, Eq. 1.14, predicts a *negative perturbation pressure*,

$$p = -\gamma \, p_0 \frac{u_0}{c}. \tag{4.5}$$

A typical particle velocity is $u_0 = 1.5$ m/s. The perturbation pressure is

$$p = -1.4 \times 100 \text{ kPa} \times \frac{1.5}{352} \approx -0.6 \text{ kPa}, \tag{4.6}$$

which explains the drop of supraglottal pressure at a glottal closure.

At step (3), with time $t = L/c$, the wavefront reaches the opening of the vocal tract. Because of the negative perturbation pressure, according to Eq. 1.43, the acoustic potential energy density in the vocal tract is

$$\mathcal{E} = \frac{1}{2\gamma p_0} p^2 = \frac{\gamma p_0}{2} \frac{u_0^2}{c^2}. \tag{4.7}$$

On the other hand, according to Eq. 1.20, the velocity of sound is

$$c^2 = \frac{\gamma p_0}{\rho_0}, \tag{4.8}$$

and then Eq. 4.7 can be converted into

$$\mathcal{E} = \frac{1}{2} \rho_0 u_0^2, \tag{4.9}$$

which equals the kinetic energy density of airflow originally in the vocal tract. Therefore, after the first phase of wavefront propagation, the aerodynamic kinetic energy of the airflow in the vocal tract immediately before a glottal closing is converted into acoustic potential energy. After the wave-front moves over the vocal tract into open air, at step (3), the particle velocity becomes zero. At that time, the air pressure in the open air is p_0, but the air pressure in the vocal tract is lower by an amount p, as shown in Eq. 4.6. Driven by that pressure difference, air particles flow from the open air into the vocal tract, see step (4). A wavefront towards the glottis is created, propagating with the velocity of sound, with a particle velocity pointing to the glottis, see step (5). After a time $t = L/c$, the wavefront reaches the glottis, see step (6). Because the glottis is now closed, the parti-cle velocity must be zero, the acoustic wavefront reflects back, see step (7), then propagates into the $+x$ direction, see step (8). This time, the particle velocity and the velocity of sound wave have the opposite direction. The air particle originally at position $x = ct$ moves to

$$x' = (c - u_0)t. \tag{4.10}$$

The volume of air mass left behind by the wavefront is reduced by a factor of $1 - u_0/c$. Therefore, the air density is changed to

$$\rho_0 \rightarrow \frac{\rho_0\, x}{x'} = \frac{ct\,\rho_0}{(c - u_0)t} \approx \rho_0 + \rho_0\frac{u_0}{c}. \tag{4.11}$$

The perturbation density of the air column is *positive*,

$$\rho = \rho_0\,\frac{u_0}{c}. \tag{4.12}$$

The air column is *compressed*, and the perturbation pressure is *positive*,

$$p = \gamma\, p_0\,\frac{u_0}{c}. \tag{4.13}$$

Nevertheless, neglecting radiation loss, the acoustic potential energy density is identical to the case of rarefied air particle, Eq. 4.7,

$$\mathcal{E} = \frac{\gamma p_0}{2}\frac{u_0^2}{c^2}. \tag{4.14}$$

After the wavefront reaches the opening of the vocal tract, see step (9), the particle velocity becomes zero. Now, the air pressure inside the vocal tract is *higher* than the pressure of open air by an amount p. Air particles flow from the vocal tract into open air to relieve the excess pressure. A wavefront propagating towards the glottis is created, see steps (10) and (11). After a time $t = L/c$, the wavefront reaches the glottis, see step (12). A condition similar to step (1) is recreated. The cycle then repeats.

The entire cycle takes four phases, each one has a time duration of L/c. The entire cycle has a period of

$$T = \frac{4L}{c}. \qquad (4.15)$$

When the wavefront reaches the open end of the vocal tract, part of the acoustic energy radiates into open air. Therefore, the acoustic wave decays with time. It creates a situation similar to the Euler transient resonator presented in Section 1.3. The waveform of the elementary wave triggered by a glottal closing is determined by the configuration of the vocal tract, representing the timbre at that moment, thus termed a *timbron*. Timbrons are elementary building blocks of human voice. Furthermore, an isolated glottal closing would trigger an isolated timbron, see Section 3.2.2. Triggered by a more or less equally timed array of glottal closings, an array of timbrons makes a vowel in the modal voice register, see Section 3.1.3. Triggered by an array of slow and somewhat irregularly timed glottal closings, an array of timbrons makes a vowel in the vocal-fry register.

4.3   Acoustic Waves in the Trachea

A glottal closing triggers a timbron in the vocal tract. It also triggers an acoustic wave in the trachea. Although the acoustic wave in the trachea is hardly perceptible by humans, it is detectable by miniature pressure sensors placed inside the trachea.

Figure 4.2 shows the acoustic process in the trachea after a glottal closing. As shown, step (0) is the state before a glottal closing, where a steady airflow runs from the trachea through the glottis into the vocal tract. After a glottal closing, see step (1), the air particle velocity at the glottis becomes zero. A zero-velocity wavefront is created and propagates downwards into the trachea. At time t, the wavefront is at $x = -ct$. Due to the particle velocity u_0, the particle initially at $x = -ct$ is moved to

$$x' = -ct + u_0 t. \qquad (4.16)$$

The volume of air column left behind by the wavefront is reduced by a factor of $1 - u_0/c$. The air density is changed to

$$\rho_0 \rightarrow \frac{\rho_0\, x}{x'} = \frac{-ct\, \rho_0}{-(c - u_0)t} \approx \rho_0 + \rho_0 \frac{u_0}{c}. \qquad (4.17)$$

The perturbation density of air column left by the wavefront is *positive*,

$$\rho = \rho_0 \frac{u_0}{c}. \qquad (4.18)$$

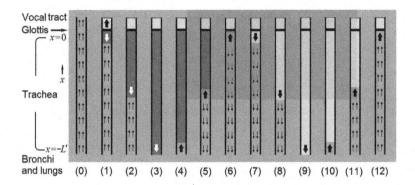

Fig. 4.2. Acoustic waves in the trachea. (0), the glottis is open, there is a uniform and steady airflow in the trachea and the vocal tract. (1) through (3), a glottal closing generates a zero-velocity d'Alembert wave front of *compressed air* propagating towards the bronchi. (4), the d'Alembert wavefront reflects by the bronchi, and the perturbation velocity of the air particles is reversed. (4) through (6), the reversed d'Alembert wavefront propagates towards the glottis. (7), the d'Alembert wavefront reflects at the glottis to generate a d'Alembert wavefront of *rarefied air*. (7) through (9), the d'Alembert wavefront of rarefied air propagates towards the bronchi. (10), the d'Alembert wavefront reflects at the bronchi to generate a d'Alembert wavefront with perturbation velocity pointing downwards. (10) through (12), the d'Alembert wavefront propagated towards the glottis to recover the initial state similar to (1).

The perturbation pressure is also positive, *with a magnitude equal to the drop of supraglottal pressure*, Eq. 4.5,

$$p = \gamma\, p_0\, \frac{u_0}{c},$$
(4.19)

which explains the surge of subglottal pressure at a glottal closing.

The bronchi and lungs have more open space than the trachea, which create a situation similar to the open end of the vocal tract. The rest of the evolution process is almost identical to the process in the vocal tract. We leave it to the readers to complete the thinking.

4.4 An Analytic Solution of the Wave Equation

In the previous sections, using intuitive arguments, we studied the acoustic waves in the vocal tract and the trachea after a glottal closing. In this section, we study the evolution of the acoustic wave after a glottal closing using the wave equation,

$$\frac{\partial^2 u(x,t)}{\partial x^2} = \frac{1}{c^2}\frac{\partial^2 u(x,t)}{\partial t^2}$$
(4.20)

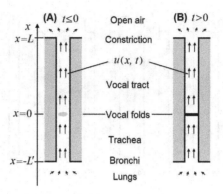

Fig. 4.3.   Initial conditions for the wave equation in the vocal tract. (A). Before a glottal closure, the air flows with an essentially uniform speed in the trachea and the vocal tract. (B) A glottal closing blocks the airflow from the trachea to the vocal tract to trigger an acoustic wave in the vocal tract and another in the trachea.

where $u(x,t)$ is the particle velocity, and c is the velocity of sound, see Fig. 4.3. Closed-form analytic solutions are obtained. That solution is derived with mathematical rigor; thus it is more convincing.

4.4.1   Initial conditions and boundary conditions

Naturally, we set the origin of time $t = 0$ at a glottal closing moment. Before the glottal closing, there is a steady airflow in the vocal tract with velocity u_0. Therefore, at $t = 0$, the initial conditions are

$$u = u_0 \qquad \text{at} \qquad 0 < x < L, \quad t = 0; \tag{4.21}$$

and because the velocity is a constant,

$$\frac{\partial u}{\partial t} = 0 \qquad \text{at} \qquad 0 < x < L, \quad t = 0. \tag{4.22}$$

At $t > 0$, the glottis is closed. The particle speed at $x = 0$ is zero,

$$u = 0 \qquad \text{at} \qquad x = 0. \tag{4.23}$$

Beyond the lips, $x > L$, there is open air. The density is a constant,

$$\rho = 0 \qquad \text{at} \qquad x > L. \tag{4.24}$$

Using the continuity equation 1.6, the boundary condition at $x = L$ is,

$$\frac{\partial u}{\partial x} = 0 \qquad \text{at} \qquad x = L. \tag{4.25}$$

The meaning of the boundary condition Eq. 4.25 is as follows: the particle density is stable in the open air; thus the particle speed is also stable.

4.4.2 Acoustic waves in the vocal tract

In this subsection, we solve the wave equation in the vocal tract after a glottal closing using Fourier series. Using the boundary conditions, Eqs. 4.24 and 4.25, we can write down the most general mathematical form of the solution using Fourier series,

$$u = \sum_{n=1}^{\infty} g_n(t) \sin \frac{(2n-1)\pi x}{2L}. \tag{4.26}$$

The correctness of the expression can be verified by direct inspection: Because each term of the sine function satisfies the boundary conditions, Eqs. 4.23 and 4.25, the sum should also satisfy such conditions. And the theory of Fourier series guarantees that the solution is the most general one. The functions of time, $g_n(t)$, are to be determined by the wave equation, Eq. 4.20, and the initial conditions, Eqs. 4.21 and 4.22.

Inserting Eq. 4.26 into the wave equation, Eq. 4.20, we find the differential equation for the functions $g_n(t)$ in Eq. 4.26,

$$\frac{d^2 g_n(t)}{dt^2} = -\left[\frac{(2n-1)\pi c}{2L}\right]^2 g_n(t). \tag{4.27}$$

The general solution is

$$g_n(t) = a_n \cos \frac{(2n-1)\pi ct}{2L} + b_n \sin \frac{(2n-1)\pi ct}{2L}, \tag{4.28}$$

where the constants a_n and b_n are to be determined by the initial conditions, Eqs. 4.21 and 4.22. Because of initial condition Eq. 4.22, $b_n = 0$. At $t = 0$, from Eqs. 4.21, 4.26 and 4.28,

$$\sum_{n=1}^{\infty} a_n \sin \frac{(2n-1)\pi x}{2L} = u_0, \qquad 0 < x < L. \tag{4.29}$$

The Fourier coefficients a_n can be easily evaluated:

$$a_n = \frac{4u_0}{(2n-1)\pi}. \tag{4.30}$$

Put the above results together, the solution is

$$u = \frac{4u_0}{\pi} \sum_{n=1}^{\infty} \frac{1}{2n-1} \cos \frac{(2n-1)\pi ct}{2L} \sin \frac{(2n-1)\pi x}{2L}. \tag{4.31}$$

Using trigonometric identity

$$\sin a \cos b = \frac{1}{2}\left[\sin(a+b) + \sin(a-b)\right] \tag{4.32}$$

and the identity [2]

$$\sum_{n=1}^{\infty} \frac{1}{2n-1} \sin\frac{(2n-1)\pi x}{2L} = \frac{\pi}{4}\,\mathrm{sgn}\,\sin\frac{\pi x}{2L}, \tag{4.33}$$

a closed-form solution is found,

$$u = \frac{u_0}{2}\left[\mathrm{sgn}\,\sin\frac{\pi(x+ct)}{2L} + \mathrm{sgn}\,\sin\frac{\pi(x-ct)}{2L}\right]. \tag{4.34}$$

The sign function $\mathrm{sgn}(x)$ in Eqs. 4.33 and 4.34 is defined as: for $x > 0$, $\mathrm{sgn}(x) = 1$; and $x < 0$, $\mathrm{sgn}(x) = -1$. Therefore, the combined function sgn $\sin(x)$ is $+1$ for $(2n-1)\pi < x < 2n\pi$, and -1 for $2n\pi < x < (2n+1)\pi$. Equation 4.34 represents a piecewise d'Alembert solution with reflections at the glottis and lips.

4.4.3   Analysis of the solution

Although the solution exists for all real values of x and t, only the values of u inside the vocal tract, $0 < x < L$, are meaningful. We will use period T in Eq. 4.15 as a scale of time. During time interval $0 < t < T/4$, the first term in the square bracket of Eq. 4.34 is always 1, and the second term is 1 for $x > ct$, and -1 for $x < ct$. In other words, $u = u_0$ for $x > ct$, and $u = 0$ for $x < ct$. Therefore, Eq. 4.34 describes a d'Alembert wavefront of zero particle velocity moving from the glottis towards the open end of the vocal tract at the velocity of sound c. Because of the equation of continuity, Eq. 1.6, behind the zero-particle-velocity d'Alembert wavefront, air is stagnant but rarefied. In fact, by integrating the equation of continuity across the wave front, the left-hand side is

$$\int_{x=ct-\epsilon}^{x=ct+\epsilon} \frac{\partial \rho}{\partial t}\,dx = \int_{x=ct-\epsilon}^{x=ct+\epsilon} \frac{\partial \rho}{\partial t}\frac{dx}{dt}\,dt = c\rho, \tag{4.35}$$

here ϵ is a small positive constant, and the right-hand side is

$$-\rho_0 \int_{x=ct-\epsilon}^{x=ct+\epsilon} \frac{\partial u}{\partial x}\,dx = -\rho_0\,u_0. \tag{4.36}$$

Therefore, the perturbation density is

$$\rho = -\frac{u_0}{c}\rho_0. \tag{4.37}$$

The air behind the wavefront is rarefied by a factor u_0/c. During the time interval $T/4 < t < T/2$, the first term in the square bracket of Eq. 4.34 is -1 for $x > c(t - T/4)$, and $+1$ for $x < c(t - T/4)$, and the second term is always $+1$. Therefore, Eq. 4.34 describes a zero-particle-velocity d'Alembert wavefront of $u = u_0$ moving from the open end of the vocal tract toward the glottis at the velocity of sound c.

During the time interval $T/2 < t < 3T/4$, the first term in the square bracket of Eq. 4.34 is always -1, and the second term is $+1$ for $x > c(t - T/2)$, and 1 for $x < c(t - T/2)$. Therefore, Eq. 4.34 describes a zero-particle-velocity d'Alembert wavefront of $u = -u_0$ moving from the glottis toward the open end of the vocal tract at the velocity of sound c. Using a similar argument in Eqs. 4.35 through 4.37, the air behind the wavefront is compressed by a factor of u_0/c.

During the time interval $3T/4 < t < T$, the first term in the square bracket of Eq. 4.34 is always $+1$, and the second term is -1 for $x > c(t - T/4)$, and $+1$ for $x < c(t - T/4)$. Therefore, Eq. 4.34 describes a zero-particle-velocity d'Alembert wavefront of $u = -u_0$ moving from the open end of the vocal tract toward the glottis at the velocity of sound c.

A complete cycle takes time $T = 4L/c$, and the inverse

$$F_1 = \frac{1}{T} = \frac{c}{4L} \tag{4.38}$$

is the *first formant frequency*. From Eq. 4.31, there is a series of formant frequencies $F_2 = 3F_1$, $F_3 = 5F_1, \ldots$, with intensities proportional to F_n^{-2}, falling off at a rate of -6 dB per octave. This fits well to the experimental observation of the vowel shwa [ə].

4.4.4 Numerical solutions for various vowels

The single-tube model of the vocal tract enables an analytic solution to make the process transparent. It approximates the production of the vowel shwa [ə]. It is apparent that by using a variable-cross-section model of the vocal tract, a numerical solution of the time-dependent wave equation should imitate the production of various vowels.

4.5   Explanations of Experimental Facts

In this Section, we will show how the experimental facts are explained by the timbron theory of voice production.

4.5.1   Superposition behavior of vowel signals

According to the timbron theory, the vowel signal is a superposition of a series of elementary waves, each starting with a stimulus, and decaying during each single pitch period. For vowels, the principle stimulus to start an elementary wave is the closing event of the glottis. After a glottal closing, the bottom of the vocal tract is closed, while the mouth and/or nose are open. The kinetic energy of the air in the vocal tract is converted into acoustic energy with a well-defined resonance characteristics. The subsequent opening of the glottis has a much smaller effect on the acoustic wave than the glottal closing event. It can bring a weak random noise to the acoustic signal, or accelerate damping by opening the vocal tract to the lungs.

4.5.2   Efficiency of voice production

For a long time, the efficiency of voice production was an unresolved puzzle. The power of breathing can be estimated as follows. From actual measurements, for example Figs. 3.16 and 3.17, the cross-glottal pressure is typically 2 kPa. On the other hand, the measurements of airflow during speaking is well documented, see Baken et al. [5]. For male speakers, the average rate of airflow is 150 cm^3/s.[1] The average power is

$$P = 2 \times 10^3 \times 150 \times 10^{-6} \approx 0.3\,\text{W}. \tag{4.39}$$

However, the average power of human voice is only 100 μW. The efficiency is about 3×10^{-5}. The efficiency gap is tremendous. The timbron theory provides insights into the problem. According to the timbron theory, only the kinetic energy of airflow inside the vocal tract immediately before a glottal closure can convert into acoustic energy. The air flow velocity in the vocal tract immediately before a glottal closing is about 1 m/s. The cross section of the pharynx is about 5 cm^2, and the average length of the pharynx is 89 mm, see Stevens [86]. Immediately before a glottal closure, the kinetic energy of airflow in the vocal tract is

$$E = \frac{1}{2}1.2 \times 1^2 \times 5 \times 8.9 \times 10^{-6} \approx 26\,\mu\text{J}. \tag{4.40}$$

At a pitch frequency of 100 Hz, the power of the kinetic energy inside the vocal tract is about 2.6 mW, and the efficiency of voice production from that power is about 4%. It is a reasonable number.

---

[1] Averaged from the measured values of mean airflow for all male speakers between age 19 to 65, Table 9-10, page 363, Baken and Orlikoff [5].

4.5.3   Role of the closed quotient

It is observed experimentally that high closed quotient corresponds to high voice intensity, and high open quotient corresponds to low voice intensity [7, 62]. The timbron theory provides a satisfactory explanation. When the glottis is closed, the vocal tract becomes a tight resonance cavity. The vocal folds are squeezed together. Because the finite thickness, moving air can hardly affect the shape of the vocal folds. Nevertheless, once the glottis opens during a resonance process, the lungs are connected to the resonance cavity. Because the lungs are porous, power decay is accelerated. The longer the open phase, the more the voice signal decays.

On the other hand, the voice intensity is not determined by the average flow rate over the entire time of a period, but the instantaneous velocity of airflow immediately before a glottal closure. If the average airflow rate is a constant, then the shorter the open phase, the higher the instantaneous air velocity immediately before a glottal closing. Because the air velocity rises slowly near the glottis opening moment, then reaches a high value before a glottal closure [4], one can describe the air velocity by a simple analytic expression such as

$$u = u_0 \sin^2 \frac{\pi(t - T_C)}{T_O}, \tag{4.41}$$

where the origin of time is the glottal closing moment, T_C is the duration of the closed phase, and T_O is the duration of the open phase. The sum is

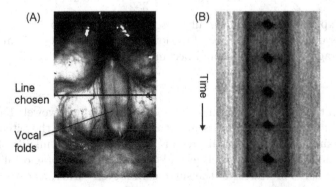

Fig. 4.4. Open quotient of an opera singer. Using videokymography, the opening and closing of an opera singer during singing are recorded. The data is more certain than using electroglottography. From the image, the closed quotient is much greater than 0.5. It is associated with a intensive and bright voice. Courtesy of Donald Miller.

the pitch period T,

$$T_C + T_O = T. \tag{4.42}$$

the relation between average glottal airflow V is

$$u_0 = \frac{2T}{T_O} \frac{V}{S} = \frac{2}{OQ} \frac{V}{S} \tag{4.43}$$

where V is the average flow rate, and S is the cross section of the pharynx. For the male speaker discussed in Section 2.2.10, for soft phonation, OQ=0.65. Take the average value of flow rate, 150 cm^3/s, and $S = 5$cm^2,

$$u_0 = \frac{2}{0.65} \frac{150}{5} \approx 90 \,\mathrm{cm/s} = 0.9 \,\mathrm{m/s}. \tag{4.44}$$

For loud phonation, OQ = 0.47. The instantaneous air velocity could become 1.3 m/s. The voice power, which is proportional to the square of the air-glow velocity, is doubled. For opera singers, the open quotient can be even smaller, and the voice is even louder, see Fig. 4.4.

4.5.4 Supraglottal pressure and subglottal pressure

The timbron theory provides a clear and quantitative explanation of the observed drop of supraglottal pressure and the surge of subglottal pressure at the glottal closing moment as well as the pressure variation during the entire pitch period. At a glottal closing moment, step (1) of Fig. 4.2, the supraglottal pressure drops. If the air velocity immediately before a glottal closure is 1.5 m/s, according to Eq. 4.6, the pressure changes by

$$p = -1.4 \times 100 \,\mathrm{kPa} \times \frac{1.5}{352} \approx -0.6 \,\mathrm{kPa}, \tag{4.45}$$

which explains the experimentally observed pressure drop.

After one half of the period of the lowest formant, see step (7) in Fig. 4.2, the supraglottal pressure should become positive,

$$p = 1.4 \times 100 \,\mathrm{kPa} \times \frac{1.5}{352} \approx 0.6 \,\mathrm{kPa}. \tag{4.46}$$

The duration of the half formant period depends on the vowel. For vowels [i] and [u], the first formant is about 300 Hz, similar to the pitch period. The maximum of supraglottal pressure takes place in the middle of a pitch period, as expected. For vowels [e] and [o], the surge of supraglottal pressure comes earlier. For vowel [a], the surge comes even earlier, as expected.

The subglottal pressure undergoes a similar cycle. As shown in Fig. 4.3, at the glottal closing moment, the subglottal pressure should surge with a identical magnitude as in Eq. 4.46. Nevertheless, the length of the trachea does not change with vowels, thus the pattern of subglottal pressure variation does not change with vowels.

4.5.5 Radiation and decay of formants

According to the detailed theoretical treatment of the radiation of acoustic waves by Phillip Morse [63], for a tube with an opening of radius a, the radiation power is proportional to the square of the ratio of the radius of the opening and the wavelength λ of the acoustic wave,

$$z_p \sim \frac{1}{2}\left(\frac{2\pi a}{\lambda}\right)^2 . \tag{4.47}$$

Therefore, if at the beginning, the resonance wave in the vocal tract has a number of components with different frequencies, the high-frequency components will radiate faster than the low-frequency components. At the beginning of a timbron, the high-frequency components are strong but they decay faster. The low-frequency components last longer.

The different decay rates of formants at different frequencies provide an adequate explanation of the observed waveforms of vowels. For vowel [i] in Fig. 3.7 and vowel [e] in Fig. 3.9, there are two groups of formants. The formant groups with frequency around 3 kHz decay at a rate of -7dB/msec. It is also apparent in the waveforms that the high-frequency signals are strongest immediately after the glottal closing, and almost disappear after a few milliseconds. The low-frequency formants, 330 Hz for [i] and 540 Hz for vowel [e], decays much more slowly, at about -2 dB per msec. This is clear in both the power curve in dB and the waveforms. The vowels [u] and [o], in Figs 3.8 and 3.10, are each dominated by one low-frequency formant, which are 400 Hz for [u], and 620 Hz for [o], decaying at -2 dB per msec. The formants of vowel [ɑ] in Fig. 3.6 have frequencies around 1 kHz. The decay rate is predictably in between, -5 dB/msec.

4.6 The Harp Analogy

It is instructive to compare vowel production and singing with a string instrument. Based on the similarity of sound, violin seems to be an obvious choice. Intuitively, the vocal folds, previously called vocal cords to hint at a similarity to violin strings, are the source of sound. The vocal tract is analogous to the top plate of the violin, which selectively amplifies the sound generated by the vocal folds. The mouth resembles the f-holes. Nevertheless, there is a paradox of size. The fundamental frequency of the violin sound is determined by the length of the string. Even the fundamental frequency of a female voice is often lower than G3, the lowest note of a violin. For a male voice, a cello or a double bass is needed to generate the fundamental frequency, requiring a string length comparable to the entire human body. However, the size of vocal folds is only about one centimeter, which is too small to function as the source of voice.

According to the timbron theory presented in Section 4.2, vowel production resembles a harp, rather than a violin. When a harpist pulls a string with a fingertip, no sound is produced. After the string is abruptly released by the fingertip, the elastic potential energy loaded on the string is converted into acoustic energy. This is similar to how an elementary acoustic wave of vowel sound is produced: During the open phase of the glottis, air flows into the vocal tract, but no sound is produced. An abrupt glottal closing starts the conversion of the kinetic energy of airflow in the vocal tract into acoustic energy. The vocal folds are not the source of sound. They constitute a valve to control the airflow. The kinetic energy of flowing air loaded in the vocal tract is the source of acoustic energy, similar to the elastic potential energy loaded by the harpist's fingertip on a string. Incidentally, the size of fingertips is approximately the size of vocal folds.

Individual notes are often played on a harp. Humans also make sounds with individual glottal closures, such as events in the vocal-fry register or glottal stops. More often, humans make repeated glottal closures with approximately equal time intervals to produce sustained vowels. The analogy in a harp is *tremolo*, where the repetition frequency of plucking is equivalent to the fundamental frequency of the vowel. There is a subtle difference between human voice organs and the harp. On a harp, tremolo cannot be played by simply plucking a single string repeatedly. A fingertip placed on a vibrating string damps it *immediately and completely*. To play a convincing tremolo, a harpist must first set the pitches of two adjacent strings to the same target pitch using pedals. By plucking the two adjacent strings alternatively, same as playing a trill, a note on one string starts before a ringing note on another string disappears. A continuous tremolo is therefore produced. In human voice organs, a glottal opening during the vibration of air in the vocal tract only causes some additional damping, but the resonance wave continues into the next pitch period.

Another subtle difference of the harp and human voice is that each string on a harp can only produce a sound of one fixed frequency. In human voice production, each configuration of the vocal tract can produce a number of frequencies, or formants. This difference is similar to the difference of the original Willis mechanism and the refined Willis mechanism proposed by Peter Ladefoged [54], as mentioned in the Preface.

Part II

Mathematical Representations

Part II

Mathematical Foundations

Part II: Mathematical Representations

The first part of the growth of a physical science consists in the discovery of a system of quantities on which its phenomena may be conceived to depend. The next stage is the discovery of the mathematical form of the relations between these quantities. After this, the science may be treated as a mathematical science, and the verification of the laws is effected by a theoretical investigation of the conditions under which certain quantities can be most accurately measured, followed by an experimental realization of these conditions, and actual measurement of the quantities.

On the Mathematical Classification of Physical Quantities

James Clerk Maxwell
Distinguished Professor of
Experimental Physics
Cambridge University

Through experiments, Michael Faraday found that electromagnetic phenomena are mediated by fields in the space, which can be visualized by using magnetic or electrical powders. James Clerk Maxwell, using partial differential equations (the mathematical tools for the theory of mechanics of continuous media) to represent the electrical and magnetic fields envisioned by Michael Faraday, resulted in a set of equations as the basis of electromagnetism as we know it.

In Part I, through analysis of experimental facts, an "atomic" theory of human voice is established. To further develop the theory of human voice, a suitable mathematical representation is needed. One would think the mathematical tools from the atomic theory of matters, that is, quantum mechanics, is applicable. This is indeed true. We will show that the concepts and mathematical tools in quantum mechanics, including principle of superposition, uncertainty relations, Laguerre functions, dispersion relations, together with Fourier analysis, can be successfully used to represent the elements of human voice, to build a concise and accurate mathematical representation.

The mathematical representation of human voice presented here is different from the legacy mathematical representations, typically using mel-frequency cepstral coefficients (MFCC) or linear-predictive coding (LPC)

[6, 25, 32, 34, 69, 70]. Those methods are asynchronous to pitch periods. A voice signal is first blocked by a shifting window, to form frames. A window function is multiplied upon each block of speech signal, such as the Hamming window. The typical frame width is 25 msec, and the typical frame shift is 10 msec. The voice signals, after being multiplied by a processing window function, undergo short-term Fourier analysis or auto-correlation analysis, frame by frame. Other properties of the framed signals are presence/absence of voicing and pitch. Each frame of voiced signal has a value of pitch, which, like the spectrum, is also treated as a slowly varying parameter of the speech signal. Such methods of signal processing are asynchronous to pitch periods. Moreover, every phoneme boundary is crossed by at least one such windows. The window functions, a necessity for using a fixed frame size, have significant adverse effects [37].

Following the timbron theory of speech production, the first step of signal processing is to extract the underlying timbrons from the signal stream, then perform Fourier analysis to the timbrons. For voiced sections of the speech signals, each timbron starts at a glottal closing moment. Therefore, for the voiced sections, the nature of the mathematical representation of human voice is *pitch-synchronous*. The advantages of pitch-synchronous signal processing methods have been known for decades, and have been tried by many researchers [44, 57, 60, 85]. However, no practical methods have been found. Chapters 5 and 6 present practical pitch-synchronous spectral analysis methods based on the timbron theory, including methods of pitch-synchronous segmentation of voice signals, Fourier analysis, and the formation of timbre vectors using Laguerre functions.

In Chapter 7, the inverse problem – to recover voice waveforms from timbre vectors, is presented. First, we show that as a consequence of the causality condition, phase spectrum of a timbron can be recovered by the dispersion relation. Strictly speaking, the application of causality condition implies that before a glottal closure, there is no voice signal. However, immediately before a glottal closure, the *glottal flow* exists, and its d'Alembert wavefront exhibits itself as a broad peak of perturbation pressure with an opposite polarity to the first sharp peak of the d'Alembert wavefront immediately following the glottal closure. We show that by adding a component in the lower-frequency part of the phase spectrum, the effect of glottal flow can be recovered. The entire voice waveform is recovered as a superposition of the timbrons.

In Chapter 8, various applications of the parameterization methods described in Chapters 5, 6, and 7 are outlined, including voice transformation, speech coding, speech recognition, and speech synthesis.

Chapter 5

Timbron Extraction

According to Part I, especially Chapter 4, voiced sound is a superposition of individual timbrons. Each timbron is triggered by a glottal closure, or for a small percentage of cases, by an incomplete glottal closure. Therefore, the problem of mathematical representation of voice is reduced to the mathematical representation of underlying timbrons. However, in most cases, a timbron starts before the previous timbrons disappear. A great majority of timbrons overlap with others. Consequently, the first step in parameterizing voice signals is to extract individual timbrons from a continuous flow of voice signal. In this chapter, various methods to extract individual timbrons in the voiced sections of speech signals are presented. Plosives can also be represented by the mathematical form of timbrons. The process of identifying and parameterizing plosives is then presented. Finally, the treatment of unvoiced sections, which is rather straightforward, is presented.

5.1   Some Mathematical Theorems

The method of timbron extraction is based on a fundamental property of timbrons, *causality*. It is a basic fact, that each timbron is an *effect* of a *cause*, a glottal closure. Before being triggered by a glottal closure, the timbron under consideration does not exist; in other words, its amplitude is zero. That simple fact has profound mathematical consequences. It is the theoretical foundation of extracting and recovering timbrons.

5.1.1   Dispersion relations

Timbrons are triggered by glottal closures. Before a glottal closure, the associated timbron does not exist. Let the time of a glottal closure be $t = 0$, a timbron $\xi(t)$ satisfies the *causality condition*

$$\xi(t) = 0 \qquad \text{for} \quad t < 0. \tag{5.1}$$

Its phase spectrum is determined by its amplitude spectrum through *dispersion relations*, or Kramers-Kronig relations according to its discoverers

[67, 97]. For a derivation of the dispersion relation, see Appendix A. Explicitly, let the Fourier transform of an timbron $\xi(t)$ be

$$F(\omega) = \frac{1}{2\pi} \int_0^\infty \xi(t)\, e^{i\omega t}\, dt. \tag{5.2}$$

The Fourier transform, in general complex, can be written in terms of an amplitude spectrum $A(\omega)$ and a phase spectrum $\phi(\omega)$ (both are real),

$$F(\omega) = A(\omega)\, e^{i\phi(\omega)}, \tag{5.3}$$

using dispersion relations, the phase spectrum can be calculated from the amplitude spectrum [67, 97],

$$\phi(\omega) = -\frac{1}{\pi} \lim_{\epsilon \to 0} \left[\int_{-\infty}^{\omega-\epsilon} \frac{\ln A(\omega')}{\omega' - \omega} d\omega' + \int_{\omega+\epsilon}^{\infty} \frac{\ln A(\omega')}{\omega' - \omega} d\omega' \right]. \tag{5.4}$$

The numerical calculation can be straightforwardly programmed on a computer, and some shortcuts can be further implemented to reduce the computation cost significantly, see Chapter 7.

The fact that a timbron is completely determined by its amplitude spectrum greatly simplifies the extraction of timbrons. We will explore that basic fact throughout this chapter and the rest of the book.

5.1.2   Uncertainty principle

For timbrons occurring at the low-pitch parts of speech, complete timbrons can often be observed, as shown in Section 3.1.3. For voices with a high pitch, only the beginning part of a timbron can be directly observed. In the following, we show that complete timbrons can be extracted from the waveform up to an accuracy limit set by the uncertainly principle.

In general, the pitch period varies constantly. For simplicity, we assume that for several periods before the period of interest, the pitch period varies slowly, such that it can be approximated as a constant. We show that for a voice generated by a series of glottal closures of pitch period T, the waveform in any complete period contains full information about the underlying timbron up to the accuracy limit set by the uncertainly principle.

Here is a proof. Because of causality condition, Eq. 5.1, an acoustic wave $X(t)$ of period T is a linear superposition of all timbrons generated by glottal closures *prior to the time of observation*,

$$X(t) = \sum_{m=0}^{\infty} \xi(t + mT). \tag{5.5}$$

Noticing that a voice signal has no constant term (i.e., no DC component), a periodic voice signal $X(t)$ can be expanded into a Fourier series,

$$X(t) = \sum_{n=1}^{\infty} \left[a_n \cos n\omega_0 t + b_n \sin n\omega_0 t \right], \tag{5.6}$$

where $\omega = 2\pi/T$, and

$$a_n = \frac{\omega_0}{\pi} \int_0^T X(t) \cos n\omega_0 t \, dt, \tag{5.7}$$

$$b_n = \frac{\omega_0}{\pi} \int_0^T X(t) \sin n\omega_0 t \, dt. \tag{5.8}$$

Using Eq. 5.5, one obtains

$$a_n = \frac{\omega_0}{\pi} \int_0^\infty \xi(t) \cos n\omega_0 t \, dt, \tag{5.9}$$

$$b_n = \frac{\omega_0}{\pi} \int_0^\infty \xi(t) \sin n\omega_0 t \, dt. \tag{5.10}$$

Therefore, a_n and b_n are real and imaginary components of the *Fourier transform* of the timbron $\xi(t)$ on interval $0 < t < \infty$ for the discrete values of frequency $\omega = n\omega_0$. The amplitude spectrum is

$$A(n\omega_0) = \sqrt{a_n^2 + b_n^2}. \tag{5.11}$$

The values of the Fourier transform are only available at integer multipliers of the value ω_0, which is $\omega_0 = 2\pi/T$. Therefore, the frequency resolution of the spectrum would not exceed the inverse of the pitch period T,

$$\Delta f \geq \frac{1}{T}, \tag{5.12}$$

which is equivalent to the uncertainly principle in quantum mechanics. As a consequence, in order to obtain a high-resolution spectrum of a vowel, the low-pitch signals are preferred. For voice signals of very high pitch, the accuracy of timbron spectrum recovery becomes low.

The above result can have an intuitive explanation as follows. In a single pitch period, the waveform is a superposition of the current timbron and the tails of previous timbrons. The Fourier analysis for a single period is equivalent to a Fourier transform over the entire time axis of a unfolded timbron. Therefore, the process is equivalent to an *unfolding* of a timbron from a folded form.

5.1.3 Independence from frame endpoints

A corollary of the previous subsections is that the amplitude spectrum obtained from the pitch-synchronous analysis is independent of the starting point within a pitch period, which only affects the phase. This point can be proved by using the exponential version of the Fourier theorem.

Let $x(t)$ be a periodic function with period T, that is,

$$x(t + T) = x(t). \tag{5.13}$$

The n-th Fourier coefficient is

$$c_n = \frac{2}{T} \int_0^T x(t) \exp\left(\frac{2n\pi i t}{T}\right). \tag{5.14}$$

By shifting the origin of the function by τ, the n-th Fourier coefficient is

$$c_n' = \frac{2}{T} \int_0^T x(t + \tau) \exp\left(\frac{2n\pi i t}{T}\right). \tag{5.15}$$

Change variable from t to $t + \tau$, we have

$$c_n' = \frac{2}{T} \int_\tau^{T+\tau} x(t) \exp\left(\frac{2n\pi i(t - \tau)}{T}\right). \tag{5.16}$$

Because of the periodicity condition, Eq. 5.13,

$$c_n' = \frac{2}{T} \int_0^T x(t) \exp\left(\frac{2n\pi i(t - \tau)}{T}\right) = c_n \exp\left(\frac{-2n\pi\tau}{T}\right). \tag{5.17}$$

Therefore, the two coefficients differ only by a phase factor. The absolute value of c_n equals the absolute value of c_n'.

This independence of the amplitude spectrum of a timbron with respect to its origin provides us freedom to choose the starting point (and also the endpoint) of a period from which the timbron is extracted.

5.2 The Ends-Matching Procedure

For the pitch-synchronous overlap add (PSOLA) method [9], the waveform of two adjacent pitch periods is multiplied by a triangular window function, and then added together. To ensure accuracy, the middle point must be set at the starting point of a pitch period where the variation is the strongest. But the end points are less important because the values of the window

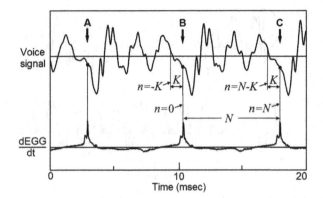

Fig. 5.1. Weakly varying section in a pitch period. The sections immediately before glottal closure, marked by A, B, and C, have weak variations than other parts in a pitch period. Taking a small section with K points to make a linear interpolation, a cyclic frame is formed. It would not notably affect the spectrum.

function near both ends are tapered to zero. For the extraction of timbrons, the requirements are different. First of all, it does not require two adjacent periods. The processing is always executed on a single pitch period. Second, because the amplitude spectrum is phase-independent, there is more freedom to select the starting point and the endpoint. However, to avoid artifacts during Fourier analysis, the values at the starting point and the value at the endpoint must be equal. Initially, in general, the values at the two endpoints of the waveform in a pitch period do not match. The acoustic wave in a pitch period must be made *cyclic* before a Fourier analysis can be performed.

As a consequence of the timbron theory, there is a natural solution to the ends-matching problem. Within each pitch period, the intensity of the voice signal decays. Near the end of each pitch period, the variation of the waveform is weak. An interpolation process can be executed in that weakly varying section to equalize the ends of the waveform without notably disturbing the amplitude spectrum.

For voice signals with simultaneous EGG signals, the peaks in the derivative of EGG signal, appearing about 1 msec before the starting of a pitch period, are the natural choice for an ends-matching process. Especially, the small segment of voice signal *before* the time of glottal closure is in the weakly varying part of a pitch period. An ends-matching procedure can be executed to make the ends match, see Fig. 5.1.

Assuming that a period BC has N points, with original PCM values $x_0(n)$, where $0 \leq n < N$. Take a small section at the end of the period with

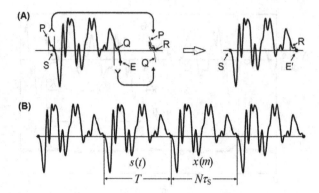

Fig. 5.2. The ends-matching procedure. (A). The raw frame starts at point S, and ends at point E. By cutting a small piece P from the previous war period, making a linear interpolation with the later part of the pitch period Q, a new piece R is formed. By substituting section Q with section R, the entire pitch period become cyclic, and the derivative is also continuous, see (B).

K points, and take another small section at the end of a previous period, also have K points. Typically $K \approx N/10$. A smoothed wave $x(n)$ is defined as follows: for $0 < n < K$, let

$$x(N - n) = x_0(N - n)\frac{n}{K} + x_0(-n)\frac{K - n}{K}; \qquad (5.18)$$

otherwise let $x(n) = x_0(n)$. Clearly, the new values of the waveform, and even their derivatives, are continuous, and the waveform is smooth. Because the new waveform is continuous and with matched ends, it is a sample of a periodic function, see Fig. 5.2. It becomes suitable for a Fourier analysis. The amended values in the weakly varying section should have insignificant effect on the spectrum.

By using pitch periods as frames and executing the ends-matching procedure, the side effects of the process windows can be completely eliminated, and the separation of timbre from pitch is complete. Furthermore, those frames are *non-overlapping*. Each PCM point only belongs to a single frame, and each frame only belongs to a single phoneme.

5.3   Segmentation Based on EGG Signals

As shown by the experimental facts that lead to the timbron theory, and implied by the effectiveness of the ends-matching procedure, to segment the speech signal into pitch periods, the exact position of the segmentation

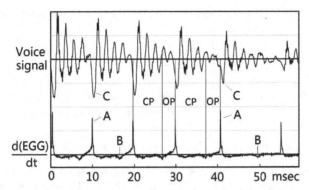

Fig. 5.3. Glottal closing moment as segmentation point. The voice signal in the open phase (OP) is much weaker than the signal in the closed phase (CP). Because the glottal closing moment is about 1 msec earlier than the starting time of the voice signal, it is always near the end of the open phase. Therefore, the glottal closing moment can be taken as the end point of the ends-matching procedure, see Fig. 5.2.

point is not critical, as long as the segmentation point is in the weakly varying part of a pitch period. In this Section, we present methods for pitch synchronous frame segmentation based on EGG signals.

5.3.1 Generating segmentation points from EGG signals

The glottal closing moments, or the position of the peak of dEGG/dt signal in a pitch period, is a good segmentation point. The glottal closing moment always precedes the starting point of the voice signal in a pitch period by a fixed interval τ, determined by the propagation time of an acoustic pulse from the glottis to the microphone, see Figs. 3.12 and 3.13. The value is typically 1 msec. Figure 5.3 shows the details of the relation. The voice signal in the open phase (OP) is much weaker than the signal in the closed phase (CP). Because the glottal closing moment is about 1 msec earlier than the starting time of the voice signal, it is always near the end of the open phase. Therefore, the glottal closing moment can be taken as the end point of the ends-matching procedure, see Fig. 5.3.

5.3.2 Estimating the average pitch period

In order to segment the entire record of a voice signal, an appropriate scale of frame duration is needed. A good time scale is the average pitch period $\overline{T}$ in the voiced section. It can be estimated as follows:

Suppose the locations of the glottal closing moments are $c_1, c_2, \ldots c_N$, where N is the number of glottal closures. The average pitch period is

$$\overline{T} = \left[\frac{1}{N-1} \sum_{n=1}^{n<N} \frac{1}{c_{n+1} - c_n} \right]^{-1}. \qquad (5.19)$$

By using the above *average of the inverse* method, the effect of abnormally large gaps between adjacent voiced sections is eliminated.

5.3.3   Voiced signals with incomplete closures

A small percentage of voiced signals can occur where the EGG signals are missing, as shown in Figs. 2.13 through 2.15. Those voiced signals are triggered by incomplete glottal closures, see Section 2.2.7. As shown in Figs. 2.13 and 2.14, the transition from periods with complete glottal closures to incomplete closures is gradual. Furthermore, only the low-frequency components are present, and often only the fundamental component identical to the pitch frequency continues to repeat. The following algorithm is found to be effective for all the five databases listed in Table 3.1.

As shown in Fig. 5.4, while the voice still goes on, the peaks in $d(\text{EGG})/dt$ signals disappear. Apparently, the vocal folds continue to vibrate, but fall short of making complete glottal closures. The glottal flow is still severely reduced, although not as sharply as those with complete closures. The closest points of incomplete closures can be estimated from speech data, see Fig. 5.4. Starting with a pitch period with complete glottal closures $P = \widehat{BC}$, take the PCM data $p(n)$ in that pitch period from $n = n_0$ to $n_0 + P$. Then, take another piece of PCM data from $n = n_0 + Q$ to $n_0 + P + Q$, where Q is a varying delay time. Define a scaling factor σ

$$\sigma = \sqrt{ \frac{\sum_{n=0}^{n<P} [p(n_0 + n)]^2}{\sum_{n=0}^{n<P} [p(n_0 + Q + n)]^2} }. \qquad (5.20)$$

Then, find the minimum of the following quantity over an interval $P/2 < Q < 3P/2$,

$$\delta = \min_{P/2 < Q < 3P/2} \frac{\sum_{n=0}^{n<P} [p(n_0 + n) - \sigma p(n_0 + Q + n)]^2}{\sum_{n=0}^{n<P} [p(n_0 + n) + \sigma p(n_0 + Q + n)]^2}. \qquad (5.21)$$

Here, the quantity δ characterizes the similarity of the two adjacent pieces of PCM data. If δ is smaller than a threshold, for example 0.05, then the value of Q at the minimum is the pitch period, or the location of the missing incomplete glottal closure. Then the process continues to find the positions of the next incomplete glottal closure until the quantity δ is no

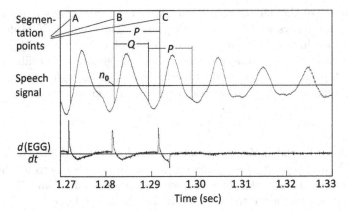

Fig. 5.4. Voiced signals with incomplete closures. The first three pitch periods of the signals have glottal closures. Segmentation points A, B, and C are derived from the $d(\text{EGG})/dt$ signals. Then the voiced signals continue, but the sharp peaks in $d(\text{EGG})/dt$ signals disappear. The instants of the incomplete glottal closures can be determined from speech signals. Source: ARCTIC databases [52], speaker jmk, sentence a0404, from 1.27 sec to 1.33 sec.

longer small. This process can also be applied to the beginning of a sequence of $d(\text{EGG})/dt$ signals. By applying this process on the ARCTIC databases, most of the incomplete glottal closures are found, and all voiced signals are segmented.

5.4 Segmentation Based on Voice Signals

The EGG signals are available only in research-type recordings. In everyday applications, only the microphone signal, or PCM, is available. In addition, huge resources of recorded speech corpora do exist, but without EGG signals. Programs to recover glottal closure instants (GCI) from voice signals are available, mostly based on LPC residual. The following method, based on the timbron theory, is simple and reliable. It was tested on the five speech corpora listed in Table 3.1, a number of audio books, and recordings using an electret microphone. Experiments showed that using the method described in this Section, the accuracy of pitch mark determination is comparable with the method based on EGG signals.

5.4.1 The profile function

According to the theory of timbrons, immediately after a glottal closure, the derivative of voice signal surges. For a given data acquisition system, the polarity of the derivative is fixed. After the surge, the derivative of the voice signal decays within each pitch period. The peak of PCM derivative near the beginning of a pitch period can be detected by multiplying the PCM derivative with an asymmetric window to generate a profile function. The peak positions of the profile function are taken as segmentation points. A simple choice of the asymmetric window function $w(n)$ is

$$w(n) = \pm \sin \frac{\pi n}{N} \left(1 + \cos \frac{\pi n}{N}\right), \quad -N < n < N, \qquad (5.22)$$

where N is the width of the asymmetric window, see Fig. 5.5. At the ends, $n = -N$ and $n = N$, the window function approaches zero as the third power of the distance to the ends, which makes it fairly smooth. The $\pm$ sign is to accommodate the polarity of the signal, see below.

To find segmentation points, the voice signal is multiplied by the asymmetric window to generate a profile function $\epsilon(m)$,

$$\epsilon(m) = \sum_{n=-N}^{n<N} [x(m+n) - x(m+n-1)]\, w(n). \qquad (5.23)$$

As shown by Eq. 5.23, it reflects the variation of the time derivative of the voice signal. An inspection to the correlation of the voice signal and the profile function reveals that the peaks A in the profile function coincide with the glottal closing moments, about 1 msec before the peak of the derivative of the voice signal, see Fig. 5.6. The section of voice signal within a few milliseconds before a peak in $\epsilon(m)$ is the section of weak variation. Therefore, the instants of the peaks in $\epsilon(m)$ are good segmentation points. However, by reversing the polarity, the peaks in $\epsilon(m)$ are located in the middle of a pitch period, see Fig. 5.6. The correct polarity of the speech signal should be determined by a test, see the following section.

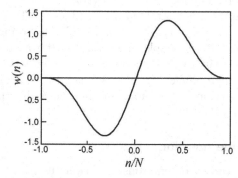

Fig. 5.5. Asymmetric window for pitch-synchronous segmentation.

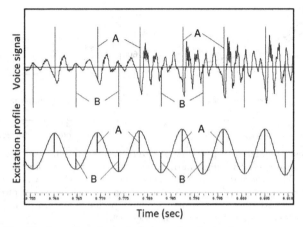

Time (sec)

Fig. 5.6. Profile function for pitch synchronous frame segmentation. By multiplying the derivative of a voice signal with an asymmetric window, a profile function is generated. If the polarity of the asymmetric window is correct, then each peak in the profile function is a few milliseconds before a surge of the derivative of the voice signal. If the polarity of the asymmetric window is reversed, then each peak in the profile function, B, is in the middle of a pitch period. The section of voice signal a few milliseconds before a peak in the profile function (with the correct polarity) represents the section of voice signal in a pitch period with weak variation.

5.4.2 Width and polarity of the asymmetric window

The width of the asymmetric window N affects the results. For a given speaker and a given data acquisition system, there is a range of optimum widths. Moreover, the polarity of the voice signal is usually unknown before processing. The following simple test can provide information of both the optimum width and the polarity of the asymmetric window.

The procedure is as follows. Take a piece of speech signal, typically a sentence. For a number of different sizes of asymmetric windows, compute the profile function and count the number of pitch marks thus produced for each size. Reverse the polarity in Eq. 5.22, and do the test again. Then plot the number of pitch marks thus produced as a function of the window size and polarity. Typical results are shown in Fig. 5.7.

In Fig. 5.7, the cases of two speakers, female slt and male speaker bdl, are shown. In both cases, (A) is the curve with correct polarity. There is a broad plateau (B) where the number of pitch marks picked up by the procedure is almost independent of the window size. With a wrong polarity, (C), false calls appear much more than the case of correct polarity. If the window is too wide, (D), many pitch marks are missed for both polarities. As shown, the plateaus of correct widths are quite broad. For slt, with a

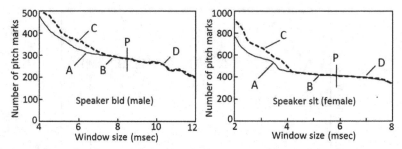

Fig. 5.7. Test the size and polarity of the asymmetric window. By segmenting a sentence with asymmetric windows of different sizes and of both polarities, the optimum window size (B) and the correct polarity (A) can be determined. Curve (C) represents the result of incorrect polarity. If the window size is too big, the number of pitch marks (D) drops quickly with window size. The average pitch period is marked by P. The optimum window size is about 0.8 times the average pitch period.

width of 4 msec to 7 msec, the result is essentially identical. For bdl, the window size could vary from 7 msec to 10 msec.

In practical implementations, for a given hardware and software, the polarity of voice signal is fixed. The width of the asymmetric window can be automatically chosen. Intuitively, the width of the asymmetric window should have a definitive relation with the average pitch period of the speaker. Experiments showed this is indeed true. For speakers bdl and slt, the average pitch period is 8.3 msec and 5.9 msec, respectively, as marked by P in Fig. 5.7. The optimum window size is approximately 0.8 times the pitch period. For a new speaker, in the beginning of the process, a universal window size, for example, 7 msec, is chosen. The running average of the pitch period is then computed periodically, for example, once each second. The window size is than updated based on the average pitch period.

5.4.3   Comparison with segmentation based on EGG

Using the asymmetric-window method, for all sentences of speakers bdl and slt of the ARCTIC databases, the pitch marks are generated, and then compared with the glottal closure instances originated from EGG signals. For speaker bdl, the window size is 7.5 msec; and for speaker slt, the window size is 5.25 msec. We use *PM* for the pitch marks generated by the asymmetric-window method, and *GCI* for the glottal-closure instances from EGG signals. A summary of the results is shown in Table 5.1.

The first part of the comparison is to count the missing PM from the GCI list, as shown in the upper half of Table 5.1. For both speakers bld and slt, more than 10% PMs are not found in the GCI list. A detailed inspection finds that most of the extra PMs are legitimate pitch marks, see Figs. 5.8

Table 5.1: Comparison of PMs and GCIs

Speaker	bdl	slt
Total number of PM	279962	411471
Missing PM in GCI	35753	47784
Percentage of missing PM	12.77	11.61
Total number of GCI	247538	368864
Missing GCI in PM	2190	4028
Percentage of missing GCI	0.88	1.09

and 5.9. Figure 5.8 is part of sentence b0413, spoken by female speaker slt, an isolated word "or". The vertical lines pointing up are pitch marks generated by the asymmetric window method. In section A, the vocal folds start to vibrate and produce several periods of acoustic wave, but there are no complete glottal closures. In Section B, the vibration of vocal folds is strong enough to generate complete glottal closures. In Section C, the vibration of vocal folds is weakened, and incomplete closures still occur. A

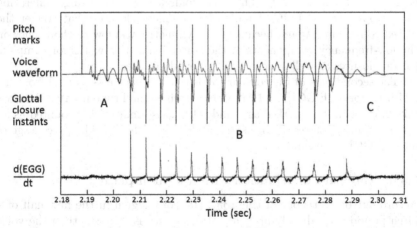

Fig. 5.8. Pitch marks without EGG signals (1). Female speaker slt, sentence b0413, an isolated word "or". In Section B, there are complete glottal closures. The PMs agree well with GCIs. In sections A and C, the incomplete glottal closures generate voice, which can be detected by the asymmetric window method, but there are no GCI signals.

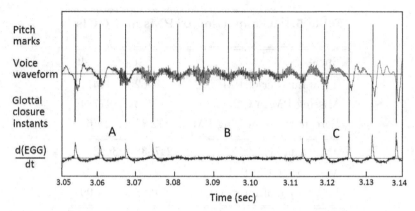

Fig. 5.9. Pitch marks without EGG signals (2). Female speaker slt, sentence b0413, part of word "pleasure". The vertical lines pointing up are pitch marks generated by the asymmetric window method. In sections A and C, the pitch marks (PM) agree well with the GCI from EGG signals. The middle section, part B, is the voiced consonant ʒ. The vocal folds are still vibrating, generating weak ripples in the EGG signal and incomplete closures trigger the voiced consonant.

few more periods of acoustic wave are produced. The asymmetric window method relies on the low-frequency components of the voice signal, which works for complete glottal closures as well as incomplete glottal closures. In Fig. 5.9, part of the word "pleasure" is shown. The middle section, part B, is the voiced consonant [ʒ]. The vocal folds are still vibrating, generating weak ripples in the EGG signal and incomplete closures that trigger the voiced consonant. Again, because the asymmetric window method relies on the low-frequency components of the voice signal, this works for complete glottal closures as well as incomplete glottal closures.

The second part of the comparison is to count the missing GCI from the PM list. For each GCI, if a PM exists within a small time interval (here we take 2.5 msec), it is marked as found. Otherwise that GCI is counted as an orphan point. The results are summarized in Table 5.1. The percentage of missing GCIs is around 1%.

5.4.4   Intensity distribution within a pitch period

According to the theory of timbrons, the voice signal in the first half of a pitch period immediately after a glottal closure is stronger than the voice signal in the second half of the pitch period. This is the basis of the segmentation process without using EGG signal, as well as the basis of the ends-matching procedure. By using the speech data in ARCTIC databases, those facts can be demonstrated directly.

For each pitch period in the voiced sections of a sentence, as segmented by the methods described in this Chapter, the PCM points in each pitch period are divided into two halves. By taking the ratio of the total power of the second halves of all pitch periods and divided by the total power of the second halves of all pitch periods, a first ratio is calculated,

$$R_1 = \frac{\sum p(n)^2|_{\text{second half}}}{\sum p(n)^2|_{\text{first half}}}, \tag{5.24}$$

which should be always smaller than 1. The histogram of the first ratios of all sentences are plotted by the solid bars in Fig. 5.10 and Fig. 5.11. The inverse of R_1, defined as

$$R_2 = \frac{\sum p(n)^2|_{\text{first half}}}{\sum p(n)^2|_{\text{second half}}}, \tag{5.25}$$

should be always greater than 1. The histogram of R_2 over all sentences are shown by the hollow bars in Fig. 5.10 and Fig. 5.11. The decay of voice intensity within a pitch period is thus well verified.

The study of the power ratio of the first half and the later half of a pitch period also provides a unambiguous method to determine the polarity of the speech signal, and consequently the polarity of the asymmetric window. If the polarity of the asymmetric window is incorrect, then the values of R_1 and R_2 would be exchanged.

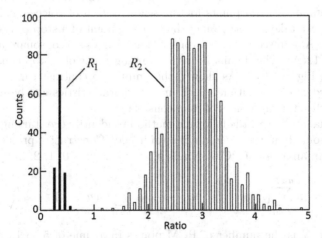

Fig. 5.10. Histograph of intensity variation in a pitch period. Male speaker bdl. The solid bars represent the histogram of the ratio R_1 in Eq. 5.24. The hollow bars represent the histogram of the ratio R_2 in Eq. 5.25. As shown, the average ratio R_1 is about 0.35, and average of R_2 is about 2.8.

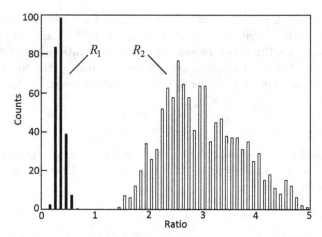

Fig. 5.11. Histograph of intensity variation in a pitch period. Female speaker slt. The solid bars represent the histogram of the ration R_1 in Eq. 5.24. The hollow bars represent the histogram of the ratio R_2 in Eq. 5.25. As shown, the average ratio R_1 is about 0.35, and average if R_2 is about 2.8.

5.5   Stops

Stop consonants, including plosives presented in Section 2.3.1 and glottal stops presented in Section 3.2.2, are very important elements of human voice. Both can be represented by timbrons. The starting points of stop consonants are critical points for the segmentation of voice signals. In the following, we take a close look of the starting point of a stop consonant, see Fig. 5.12. A 30-msec section of voice signal with a stop consonant are shown in Fig. 5.12(A). The details near the starting point of a stop consonant in shown in Fig. 5.12(B). As shown, the amplitude contrast can be as large as two orders of magnitude. Such unique characteristics can be utilized to identify the starting points of stop consonants.

By trial and error, the following method to identify the starting points of stop consonants in voice signals is established. Construct a profile function $S(m)$ from an array of PCM signals $x(n)$, where $n = 0, 1, 2, \ldots$ by

$$S(m) = \ln \sum_{n=0}^{n<N} [x(m+n)]^2 \cos \frac{n\pi}{N} - \ln \sum_{n=0}^{n<N} [x(m-n)]^2 \cos \frac{n\pi}{N}. \quad (5.26)$$

The limit N is the number of PCM points in a time of 5 to 10 msec. By looking into Fig. 5.12, it is clear that before a stop consonant begins, the profile function gradually increases to a high plateau. Immediately after the stop consonant starts, the profile function abruptly drops. Therefore, the

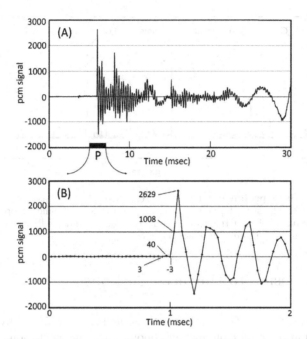

Fig. 5.12. Waveform of a stop. (A), A 30-msec section of voice signal with a stop consonant [k]. The characteristic oscillation could last more than 10 msec. (B), details near the starting point of a stop consonant. As shown, the amplitude contrast can be as large as two orders of magnitude. Using a computer program, the starting point of a stop consonant can be identified.

negative derivative of $S(m)$ over m

$$P(m) = S(m) - S(m+1) \qquad (5.27)$$

should have a sharp peak at the starting point of a stop consonant. Consequently, the peaks in the profile function $P(m)$ are the starting points of stop consonants.

Figure 5.13 shows an example of the profile function of a complete sentence. As shown, the value of the profile function at the starting point of a stop consonant can be as high as a few thousands, while the background amplitude of $P(m)$ is below a few hundreds. Therefore, using a threshold anywhere between 1000 and 2000 would pick up the starting points of stop consonants. By looking at the PCM signals and the stop markings, it appears that the great majority of stops are detected. The inclusion of the starting points of the stop consonants improves the parameterization and regeneration of speech signals.

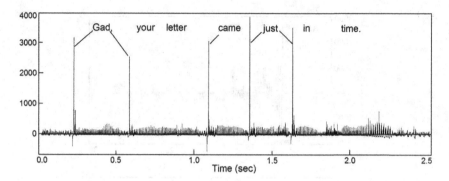

Fig. 5.13.   Stop profile function. Source: ARCTIC database [52], speaker bdl, sentence a0008. The stops [g] and [dʒ] in "Gad", stop [k] in "come", stops [dʒ] and [t] in "just", are detected. But the [t] in "letter" is a flap [ɾ], and the [t] in "time" is in the shadow of [n] in "in", therefore are not detected.

5.6   Extension to Unvoiced Sections

The pitch marks and the starting points of stop consonants are the basic segmentation points for pitch-synchronous frame segmentation. In the entire record of speech signals, there are many gaps between the ends of voiced sections, and between those ends and the starting points of stop consonants. To process the entire speech signal, the gaps should also be segmented.

To insert segmentation points into the gaps between the voiced groups and stops, each of the time gaps is divided by the average pitch period defined by Eq. 5.19, and the integer part of that quotient is taken as the number of additional segmentation points for that time gap. By dividing the number of PCM points in the gap by the number of segmentation points, and taking the integer, a step size for that gap is obtained. Using that step

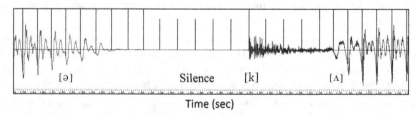

Time (sec)

Fig. 5.14. Segmentation including stops. The long vertical lines are starting points of pitch periods of vowels, produced by the asymmetric window method, and the starting points of stop consonants, produced by the stop identification procedure. The short vertical lines are unvoiced consonants and silence, produced by the extension process.

size, locations of additional segmentation points are determined. Those segmentation points are almost equal-distanced, with a size close to that of the pitch periods in the voiced sections.

The pitch marks generated by the asymmetric window approach are equivalent to the pitch marks originated from EGG signals. By combining the starting points of stops and extending into the gaps, including unvoiced consonants and silence, the entire speech signal is segmented into pitch-synchronous frames. Figure 5.14 is an example. The vertical lines are segmentation points. The long vertical lines are starting points of pitch periods of vowels, produced by the asymmetric window method, and the starting points of stop consonants, produced by the stop identification procedure. The short vertical lines are unvoiced consonants and silence, produced by the extension process in Subsection 5.6.

5.7 Amplitude Spectrum of a Timbron

Spectral analysis is a basic tool to study speech and voice signals. The traditional method of spectral analysis is first to block the speech signal into overlapping frames with a fixed frame size and a fixed frame shift, then multiply each frame with a processing window. Typically, fast Fourier transform (FFT) is applied to generate the amplitude spectrum.

The spectral information obtained by the fixed windows depends on the size of the windows, as shown in Fig. 5.15. (A) is the spectrogram with a 32-msec processing window, *longer* than a pitch period. The main feature in the spectrum is the pitch frequency and its overtones. The temporal resolution of the spectrum is roughly the size of the processing window, 32

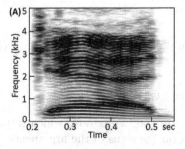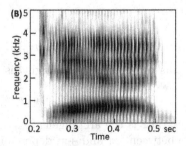

Fig. 5.15. Pitch-asynchronous spectral analysis. (A), spectrogram processed with a 1024-point window, 32 msec. The spectrum is dominated by overtones of the fundamental frequency of pitch. (B), spectrogram processed with a 128-point window, 4 msec. The spectrum depends on the position in each pitch period. Source: ARCTIC database [52], speaker bdl, sentence a0008. The word "Gad".

msec. (B) is the spectrogram with a 4-msec processing window, *shorter* than a pitch period. As shown, the power spectrum depends the timing of the window. For frames centered at different phases within a pitch period, the power spectrum can be vastly different, as shown by the vertical lines in the spectrogram.

5.7.1 Spectral analysis of timbrons

Using the pitch-synchronous segmentation and ends-matching methods, the spectral information of the timbrons, which is *completely separated from the pitch information*, is obtained. As shown in Fig. 5.2(B), for a cyclic piece of signal $s(t)$, Fourier theorem states that the signal can be accurately represented by a Fourier series,

$$s(t) = \sum_{n=1}^{\infty} \left[a_n \cos \frac{2n\pi t}{T} + b_n \sin \frac{2n\pi t}{T} \right], \tag{5.28}$$

where the Fourier coefficients are

$$a_n = \frac{2}{T} \int_0^T s(t) \cos \frac{2n\pi t}{T}, \tag{5.29}$$

and

$$b_n = \frac{2}{T} \int_0^T s(t) \sin \frac{2n\pi t}{T}. \tag{5.30}$$

Because of the capacitance in the circuit, the DC component of a microphone voice signal is always zero. The amplitude spectrum $A(f)$ of the signal, at the integer multiples of the pitch frequency f_0 defined by

$$f_0 = \frac{1}{T} \tag{5.31}$$

is

$$A(nf_0) = \sqrt{a_n^2 + b_n^2}. \tag{5.32}$$

The phase spectrum of the signal is

$$\phi(nf_0) = \text{atan2}(b_n, a_n), \tag{5.33}$$

where atan2 is the arctangent function in the C program, which returns a phase between $-\pi$ and $+\pi$, depending on the signs of the arguments.

Practically, the voice signal is digitized with a sampling frequency f_S, or a sampling interval τ_S, which is the inverse of sampling frequency,

$$\tau_S = \frac{1}{f_S}. \tag{5.34}$$

For example, at a sampling rate of $f_S = 44.1$ kHz, the sampling interval is $\tau_S = 0.022675$ msec. Denote the digitized signal $x(m)$ as

$$x(m) = s(m\tau_S), \qquad (5.35)$$

and notice that the number of points in a period is

$$N = \frac{T}{\tau_S}, \qquad (5.36)$$

the Fourier integral can be approximated by a finite sum,

$$a_n = \frac{2}{T} \sum_{m=0}^{m<N} \left[x(m) \cos \frac{2nm\tau_S\pi}{T} \right] \tau_S = \frac{2}{N} \sum_{m=0}^{m<N} \left[x(m) \cos \frac{2nm\pi}{N} \right], \qquad (5.37)$$

and similarly,

$$b_n = \frac{2}{N} \sum_{m=0}^{m<N} x(m) \sin \frac{2nm\pi}{N}. \qquad (5.38)$$

According to the Nyquist theorem, the maximum number of n is the integer part of $[N/2]$. The Fourier's theorem is approximately valid,

$$x(m) = \sum_{n=1}^{n<[N/2]} \left[a(n) \cos \frac{2nm\pi}{N} + b(n) \sin \frac{2nm\pi}{N} \right]. \qquad (5.39)$$

The amplitude spectrum of the periodic signal is still represented by Eq. 5.32, and the phase spectrum is still represented by Eq. 5.33, but the number of points is limited by the Nyquist condition $n < [N/2]$.

The amplitude spectrum and the phase spectrum thus computed only exist at the integer multiples of the pitch frequency. Because the pitch frequency of each period is different, such a representation is inconvenient for comparison. A general procedure to interpolate to a standard scale of frequency values is needed. By trial and error, it was found that the truncated Whittaker-Shannon interpolation produces consistent results. While the original spectrum is only defined at the points $f = nf_0$, the Whittaker-Shannon interpolation generates values of the amplitude spectrum at any point of the frequency axis with a brick-wall low-pass filter,

$$A(f) = \sum_{n=-\infty}^{\infty} A(nf_0) \, \text{sinc} \left(\frac{f - nf_0}{f_0} \right), \qquad (5.40)$$

where the sinc function, shown in Fig. 5.16(A), is defined as

$$\text{sinc}(x) = \frac{\sin \pi x}{\pi x}. \qquad (5.41)$$

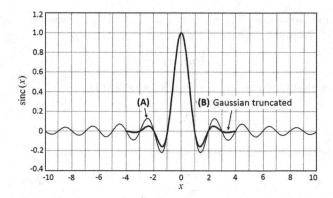

Fig. 5.16. The sinc function, original and truncated. (A), the original sinc function. (B), truncated by a Gaussian function to eliminate the oscillation in the interpolated spectrum and to speed up computation, see Eq. 5.42.

Its value at $x = 0$ is determined by continuity condition, $\mathrm{sinc}(0) = 1$. However, it creates too much oscillation in the interpolated spectral curve, and the computation is slow. This can be resolved by truncating the sinc function with a Gaussian profile, see Fig. 5.16(B),

$$\mathrm{Sinc}(x) = e^{-\kappa x^2} \frac{\sin \pi x}{\pi x}, \qquad -4 < x < 4, \tag{5.42}$$

where κ is optimized by experimentation. A good value is $\kappa = 0.16$. The excess oscillation is eliminated, and the computation is accelerated. Take n_0 be the integer part of f/f_0, the sum only takes a few terms,

$$A(f) = \sum_{n=-4}^{4} A((n_0 + n)f_0) \, \mathrm{Sinc}\left(\frac{f - (n_0 + n)f_0}{f_0}\right). \tag{5.43}$$

To facilitate numerical computation, the nominally continuous amplitude spectrum is still represented by discrete frequency values. A convenient choice is 512 points on a 22.05 kHz scale. Each frequency interval is 43.066 Hz, which is fine enough to represent timbre phenomena.

Examples of amplitude spectra of three vowels and one consonant spoken by four speakers are shown in the following section.

5.7.2 Examples

In section 3.1.3, while presenting the PCM signals of vowels, we have briefly touched the amplitude spectra of vowels by a single speaker, where several

Table 5.2: Formants of vowel [ɑ] by four speakers

Speaker	(F_0)	F_1	F_2	F_3	F_4
bdl (male, US English)	210	**710**	**1190**	2610	3200
ocm (male, Mandarin)	130	**740**	**1180**	2850	
slt (female, US English)	170	**860**	**1360**	2860	4100
ocf (female, Mandarin)	310	**1210**	**1640**	2600	3160

consecutive pitch periods of each vowel are analyzed. In this section, amplitude spectra of three vowels plus one consonant spoken by four speakers are analyzed and compared. As shown, because the pitch information and timbre information are completely separated, great details of the amplitude spectrum characteristic to the timbre are displayed. Figure 5.17 shows the amplitude spectra of vowel [ɑ], spoken by two male speakers and two female speakers. The peak frequencies are listed in Table 5.2. The most prominent feature is the two strong peaks near 1 kHz, marked by bold-faced numbers. Female speakers have higher frequencies than male speakers. Within each gender, the peak frequencies can also be significantly different, reflecting individual character. There are several weak peaks around 3 kHz. The de-

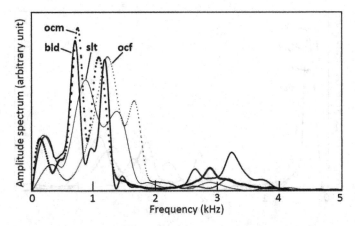

Fig. 5.17. Amplitude spectra of vowel [ɑ]. The general features are the two strong peaks around 1 kHz. The frequencies of those peaks depend on the speaker. The peaks around 3 kHz are much weaker and vary significantly.

Table 5.3: Formants of vowel [i] by four speakers

Speaker	F_1	F_2	F_3	F_4
bdl (male, US English)	310	2530	3080	3640
ocm (male, Mandarin)	300	2030	2880	3510
slt (female, US English)	260	2800	3110	3460
ocf (female, Mandarin)	300	3120	3410	3600

tails of the frequency and intensity depend on the speaker. There is also a peak at the very low frequency end, marked here as the zeroth formant. It is related to the pitch frequency. The origin of that peak will be discussed in Chapter 7.

Figure 5.18 shows the amplitude spectra of vowel [i]. The frequencies of the peaks are listed in Table 5.3. The dominant features are a very strong peak at about 300 Hz, and a group of medium-intensity peaks between 2 kHz and 4 kHz. The low-frequency peak is almost identical for all speakers, but the high-frequency features are strongly speaker-dependent. It is hard to say exactly how many peaks are in the high-frequency group, because the actual shape of the amplitude spectrum is rather complicated. Those peaks are probably affected by the detailed geometry of the elements in the frontal oral cavity, including the tongue, the teeth, and the lips.

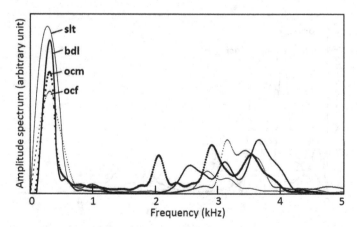

Fig. 5.18. Amplitude spectra of vowel [i]. The general features are one strong peak at a low frequency near 300 Hz, and a group of medium-intensity peaks in the range of 2kHz to 4kHz, with details depending on speaker.

Table 5.4: Formants of vowel [u] by four speakers

Speaker	F_1	F_2
bdl (male, US English)	**330**	(670)
ocm (male, Mandarin)	**410**	(690)
slt (female, US English)	**280**	(3700)
ocf (female, Mandarin)	**410**	(740)

Figure 5.19 shows the amplitude spectra of vowel [u]. The frequencies of the peaks are listed in Table 5.4. As shown, the only universal feature is a strong peak at a low frequency around 300 and 400 Hz. The additional features are very weak and varying. Because of the sharp intensity ratio between the main peak and the secondary features, it is likely that the amplitude spectrum is dominated by the main peak. The secondary features have an insignificant effect on the voice. Rather, the detailed shape of the main peak should have a greater effect to the voice.

Here, we should ask a general question about an adequate description of the amplitude spectra of vowels. Although the concept of formants makes good intuitive sense, the observed spectra cannot be represented simply by listing a fixed number of formants, for example, 3 or 4. In the case of vowel

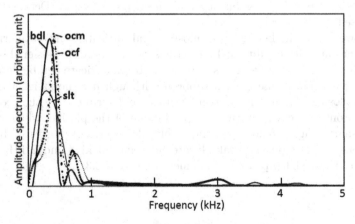

Fig. 5.19. Amplitude spectra of vowel [u]. The only universal feature is a strong peak at a low frequency around 300 and 400 Hz. The additional features are very weak and varying.

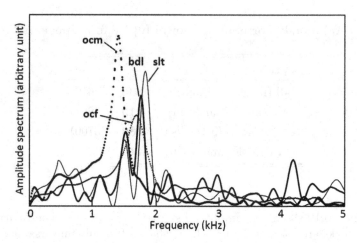

Fig. 5.20. Amplitude spectra of consonant [k]. There is a sharp peak between 1.5 kHz and 3 kHz, representing the resonance in the front oral cavity, see also Figs. 2.17 and 2.19.

[u], a single peak in the spectrum is probably sufficient, but the shape of the single peak should be parameterized with great care.

In Section 2.3.2, especially Fig. 2.22, the amplitude spectra of fricatives were discussed. The frames are not synchronous with glottal closures, but are segmented with a frame size similar to the average pitch period. The main features are in the high end of the frequency scale, and noisy: not as regular as those of the vowels. To represent the spectra, the resolution in the high end of the frequency scale need not be high. Details will be discussed in Chapter 6.

However, at the beginning of plosives and affricates, the waveform satisfies the causality condition: Before the starting instant, the signal is zero. Therefore, the starting features of a plosive or an affricate can be described by the same mathematics of a timbron, although it is not triggered by a glottal closing. The main spectral features are determined by the geometry of the front oral cavity during the production of the plosive, which can be quite sharp. Figure 5.20 shows the amplitude spectra of plosive [k] by four speakers. The strongest peaks locate between 1.3 kHz and 2 kHz. The origin of those sharp peaks is explained in Section 2.3.1.

Chapter 6

Timbre Vectors

As we have presented in Chapter 5, for voiced sections of speech signals, the amplitude spectrum of each pitch period defines the underlying timbron. However, the amplitude spectrum is inconvenient and over-specified. For vowels, the features of the amplitude spectrum concentrate in the low-frequency range, typically 0.3 kHz to 5 kHz. A high frequency resolution is required only in that frequency range. For fricatives, the amplitude spectrum is mostly in the high-frequency range, but the required frequency resolution is low. In this Chapter, we introduce a mathematical representation of the timbre spectrum, the timbre vector, which follows the human sensitivity of frequency. The timbre vector can be viewed as a vector of unit norm in a Hilbert space, resembling a state vector in quantum mechanics. Thus the mathematical tools in quantum mechanics can be applied.

6.1 Distortion Measures

In speech recognition and speech coding, an essential quantity to be defined is the *distortion measure* or simply *distance* between two speech samples [71]. It is the measure of dissimilarity of the two speech samples. The definition of distance should satisfy the following conditions:

First, the distance should be phonetically relevant. If a listener found that the two pieces of speech signal are similar, the distance should be smaller. On the other hand, if two pieces of speech signals are phonetically different, the distance should be greater. Apparently, raw waveforms do not create a good distance. And the difficulties in defining a distance based on LPC (linear prediction coding) are well known [71].

Second, the distance d for all elements $\mathbf{x}$, $\mathbf{y}$ and $\mathbf{z}$ in the space of consideration should satisfy the following conditions:

- **positive definitiveness**
  $0 \le d(\mathbf{x}, \mathbf{y}) \le \infty$, and $d(\mathbf{x}, \mathbf{y}) = 0$ if and only if $\mathbf{x} = \mathbf{y}$.

- **symmetry**
  $d(\mathbf{x}, \mathbf{y}) = d(\mathbf{y}, \mathbf{x})$.

- **triangle inequality**
  $d(\mathbf{x}, \mathbf{y}) \leq d(\mathbf{x}, \mathbf{z}) + d(\mathbf{z}, \mathbf{y})$.

One of the criteria for choosing a speech parameterization method is the ability to define a good distortion measures. In speech recognition, the most widely used parameterization is that of mel-frequency cepstral coefficients, which has a straightforward definition of distortion measure.

6.1.1   Mel-frequency cepstral coefficients

Since its introduction in early 1980s [24], mel-frequency cepstral coefficients (MFCC) have become a popular parametrical representation in speech recognition. The process of building MFCC is as follows:

1. Block the speech signals with a shifting frame, multiply it by a window function, see Section 8.3.2.

2. Take the amplitude spectrum of a windowed frame of speech signal using fast Fourier transform (FFT).

3. Using a set of triangular windows to map it into the mel scale.

4. Take the logarithm of the binned amplitude spectrum.

5. Take the discrete cosine transform of the logarithms.

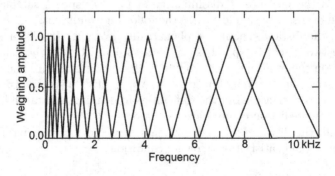

Fig. 6.1.   Mel-scaled triangular windows. The human perception of frequency scale is non-linear. To build MFCC, a set of rectangular windows of non-linear scale is multiplied to the spectrum, to generate a set of coefficients. Since the number if coefficients is limited to 10 to 15, the accuracy is compromised. A better method is to use Laguerre functions to represent the non-linear frequency scale. See Section 6.2.

The third step, binning the amplitude spectrum using a mel-scaled triangular windows, is shown in Fig. 6.1. Human perception of frequency scale is not linear to the frequency scale. Humans are more sensitive to frequency differences in the low-frequency range, represented by the mel scale [24]. Therefore, a non-linear distribution of the bins would accommodate the non-linearity of human perception on the frequency scale. Second, as shown in Fig 5.15 (A), if the size of the processing window is greater than a pitch period, the fine features in the amplitude spectrum are the pitch frequency and its overtones. Using the bins, the peaks and valleys of the pitch frequency and its overtones are smoothed out.

The fourth step, taking the logarithm, is to take care of the intensity problem. This further reduces the effect of pitch frequency.

The last step, discrete cosine transform, further reduces the dimension of the output vector, and makes it easier for the definition of a distortion measure, which is simply the Euclidian distance [71],

$$d = \sqrt{\sum_{n=0}^{n<N} (c_n - c'_n)^2}, \qquad (6.1)$$

where c_n and c'_n are the n-th MFCC coefficients of two frames, and N is the dimension of the MFCC coefficients, typically $N = 13$. Because the coefficients are derived from a discrete cosine transform, the definition satisfies the three mathematical requirements stated above.

However, by reducing the original spectrum into cepstral coefficients, a lot of fine detail is lost. The original dimension of an amplitude spectrum could be in the hundreds. Therefore, the original amplitude spectrum can never be recovered with an appropriate accuracy. The step of taking the logarithm also generates some trouble, because the noise level in the low-power sections is excessively amplified. Especially, the MFCCs in the silence sections are full of nonsense signals originated from low-energy noise, which must be handled by a special "silence modeling".

6.1.2   Distortion measures for LPC coefficients

Because the LPC coefficients are very sensitive to the fine details of the data, and do not form a Euclidean space, the definition of distorsion measures of LPC coefficients is a difficult and tricky problem. For details, see [71].

6.2   Timbre Vectors

In Chapter 5, a systematic method to extract timbrons from speech signals, especially voiced signals, is presented. Since the phase spectrum of a timbron is uniquely determined by its amplitude spectrum, the amplitude spectrum fully defines a timbron. And the spectra of timbrons are completely separated from pitch information.

However, direct use of the amplitude spectrum on the frequency scale is not convenient and does not follow the human perception on the frequency scale. The method of binning in the formation of MFCC, presented in Section 6.1.1, is too crude and will cause a serious loss of information. In this section. we show that using a standard method in least squares approximations with a *weight function* chosen to represent the human perception of frequency [1, 2], a more convenient and more accurate mathematical representation of human voice can be constructed, and well-behaved distortion measures, the *timbre distance*, can be straightforwardly defined.

6.2.1   Laguerre functions

According to the experimental results of human perception on the frequency scale, see Fig. 1.10, human ears are most sensitive to signals in the frequency range around 0.3 kHz to 5 kHz. The peak of human sensitivity is around 3 kHz. For signals of very low frequency, for example lower than 100 Hz, human sensitivity is very low. For signals of frequency higher than 10 kHz, the sensitivity and frequency resolution are also very low. To better represent the human sensitivity to frequency, a good analytic form of a *weight function* [1, 2] is

$$w(x) = x^k e^{-\kappa x}, \tag{6.2}$$

where k is a positive integer, and κ is a scaling constant. A graph of that weight function with $k = 4$ is shown in Fig. 6.2. As shown, high resolution of that weight function is found in the frequency interval of 0.3 kHz to 5 kHz, with a peak resolution around 3 kHz. The weight function is a good approximation of human sensitivity in frequency.

A classical mathematical problem is to find a set of orthogonal polynomials $P_n(x)$ in the interval $0 \leq x < \infty$ satisfying the following condition,

$$\int_0^\infty w(x) P_m(x) P_n(x) = \delta_{mn}, \tag{6.3}$$

where δ_{mn} is the Kronecker delta,

$$\delta_{mn} \equiv \begin{cases} 0 & \text{if } m \neq n, \\ 1 & \text{if } m = n. \end{cases} \tag{6.4}$$

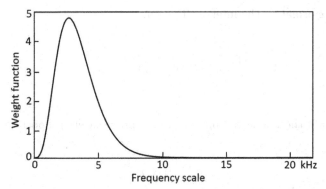

Fig. 6.2. Weight function for Laguerre polynomials. A weight function of Laguerre polynomials of rank 4 on a 22.05 kHz frequency scale. The highest frequency resolution is in 0.3 kHz to 5 kHz, where most of the formants are located.

The standard solution of this problem is found in the *Laguerre polynomials*, well-known as the mathematical forms of the wavefunctions of the hydrogen atom in non-relativistic quantum mechanics [2].

For the mathematical details of the Laguerre polynomials, see Appendix B. Here we discuss some properties of the Laguerre polynomials, mostly using graphics. The Laguerre polynomials are defined by the Rodriguez formula,

$$L_n^{(k)}(x) = \frac{e^x}{n!x^k}\frac{d^n}{dx^n}(e^{-x}x^{n+k}),\tag{6.5}$$

where n and k are non-negative integers. Explicitly, the first three Laguerre polynomials are

$$L_0^{(k)}(x) = 1,\tag{6.6}$$

$$L_1^{(k)}(x) = k - 1 - x,\tag{6.7}$$

and

$$L_2^{(k)}(x) = \frac{x^2}{2} - (k+2)x + \frac{(k+2)(k+1)}{2}.\tag{6.8}$$

In general,

$$L_n^{(k)}(x) = \sum_{m=0}^{n}(-1)^m\frac{(n+k)!}{(n-m)!(n+m)!m!}x^m.\tag{6.9}$$

The Laguerre functions

$$\Phi_n^{(k)}(x) = \sqrt{\frac{n!}{(n+k)!}}\ e^{-x/2}x^{k/2}L_n^{(k)}(x)\tag{6.10}$$

are orthonormal on the interval $(0, \infty)$,

$$\int_0^\infty \Phi_m^{(k)}(x)\,\Phi_n^{(k)}(x)\,dx = \delta_{m,n}. \qquad (6.11)$$

Near $x = 0$, the Laguerre functions are approximately proportional to

$$\Phi_n^{(k)}(x) \propto x^{k/2}. \qquad (6.12)$$

Because the amplitude spectrum is an even function of frequency, only for $k = 0$, $k = 4$, ... the Laguerre functions are analytic even functions of x. Because near $x = 0$, $\Phi_n^{(0)}(x)$ always equals to 1, and that values oscillates quickly, $\Phi_n^{(4)}(x)$ is a good choice.

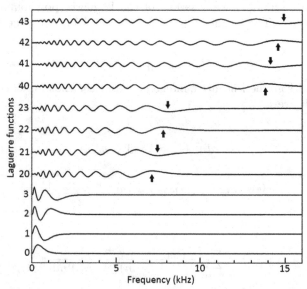

Fig. 6.3. Waveforms of Laguerre functions. Laguerre functions of selected order n are displayed. For Laguerre functions of lower n, the values concentrate in the low frequency range. For Laguerre functions of higher n, the values concentrate in the high frequency range, and the strongest peaks are near the high-frequency end of the waveform, with alternative sign.

6.2.2 Definition of timbre vector

The amplitude spectrum of a timbron $A(\omega)$ can be approximated by a sum of N such Laguerre functions,

$$A(\omega) \cong \sum_{n=0}^{n<N} C_n \Phi_n(\kappa\omega),\qquad(6.13)$$

where the coefficients C_n are

$$C_n = \int_0^{\infty} \kappa A(\omega)\Phi_n(\kappa\omega)\,d\omega.\qquad(6.14)$$

The scaling factor κ is chosen to optimize accuracy.

The coefficients C_n form a vector, denoted as $\mathbf{C}$. The norm of the vector

$$\alpha \equiv \|\,\mathbf{C}\,\| = \sqrt{\sum_{n=0}^{n<N} C_n^2}\qquad(6.15)$$

represents the overall amplitude of the period. The *normalized Laguerre spectral coefficients*, constitute a *timbre vector*

$$\mathbf{c} = \frac{\mathbf{C}}{\|\,\mathbf{C}\,\|},\qquad(6.16)$$

represents the spectral distribution of the period, characterizing the timbre of that pitch period independent of duration and intensity.

6.2.3 Feature vector of a frame

A complete feature vector of a frame, which is a pitch period for voiced signals, also contains a voicedness index, a pitch period, and an intensity parameter, which is the root-mean-square (rms) value of the PCM in the frame. We will simply call it a *feature vector*:

υ	voicedness index
τ	pitch period in msec
α	rms amplitude in PCM units
$\mathbf{c}$	timbre vector

Directly from the segmentation procedure described in Section 5.3 or Section 5.4, the voicedness index is binary. However, for voiced fricatives and the transitional frames between a vowel and an unoiced fricative, *fractional voicedness index* can be assigned and implemented in the waveform recovery process, see Section 7.4.2. The fractional voicedness indices can improve the smoothness and naturalness of reproduced voice.

The format of the feature vectors used in this book typically has 64 double-precision floating-point numbers. Therefore, the dimension of the timbre vector is 61. In a certain sense, this is an overkill. However, during computation, using the state-of-the-art computer systems, it is not a problem. For telecommunication, using vector quantization, the coded version can be reduced to small integers, often smaller than 4096, or 12 bits. The large dimension of the timbre vectors would improve the quality of decoded speech, but poses no restriction to the bandwidth in telecommunication; see Chapter 8 for details.

6.3   Timbre Vector and Amplitude Spectrum

One of the advantages of using Laguerre functions is that its absolute convergence, thus the dimension of the timbre vector is scalable. For MFCC and LPC, the number of coefficients is typically 13. Including the derivatives and second derivatives, the number is 39. By changing the dimension, the values would be completely different. The number of Laguerre functions can be much greater, as in this book, where we use a 61-dimension timbre vector. Nevertheless, one can truncate it to any smaller dimension as needed. In other words, the timbre vectors are scalable.

Figure 6.3 shows the waveforms of Laguerre functions with different n. As shown, for Laguerre functions of lower n, the values concentrate in the low frequency range. The waveform of $\Phi_0(x)$, the lowest curve in Fig. 6.3, seems to coincide with a typical first formant. The subsequent Laguerre functions make refinements to the amplitude spectrum in the low-frequency range. The Laguerre functions of middle range of n have strongest peaks in the middle range of the frequency scale. The Laguerre functions of large n have broad and strong peaks at the upper end of the frequency scale. Those Laguerre functions may refine the spectrum of vowels, but the effect is not large. One can expect that for vowels, the coefficients of lower n are dominant, while for fricatives, the coefficients of higher n are dominant. This is indeed the case.

The Laguerre polynomials are optimized for least-squares approximations with a weight function Eq. 6.2 and Fig. 6.2. Therefore, we would expect that the amplitude spectrum recovered from such a Laguerre expansion is more accurate in the low-frequency range, and less accurate in the high-frequency range. In the following, we will show a number of examples for vowels and unvoiced consonants from the first sentence a0001 of the ARCTIC databases, spoken by a male speaker bdl. The spectrogram with the phonetic transcription of that sentence is shown in Fig. 6.4. The text is

Author of the danger trail, Philip Steels, (etc.)

Figure 6.5 shows, for vowels [ɔ] and [ə], the original amplitude spectra, displayed by dotted curves; amplitude spectra recovered from the timbre vectors, displayed by thin solid curves; and a plot of the values of the timbre vectors with the index of the timbre vector indices as the x-axis. As shown, the timbre vectors represent the amplitude spectra in high fidelity with fine details. The elements in the timbre vectors also form continuous curves. The timbre-vector elements of lower indices, from 0 to 10, represent the gross features of the amplitude spectra; whereas the elements of higher indices, from 10 to 45, represent the fine details of the amplitude spectra. This point will become clear later, especially for the unvoiced consonants and voiced consonants.

Figure 6.6 shows the same features as in Fig. 6.5 for vowels [i] and [e], respectively. Again, as shown, the fine details of the original amplitude spectra are reproduced in high accuracy by the timbre-vector representation. Notice here that the dimension of timbre vectors is 45, but the amplitude spectrum has 372 points, with a frequency resolution of 43 Hz.

Figures 6.7 and 6.8 are amplitude spectra of four unvoiced consonants. As we have discussed in Chapter 3, those are noise sounds, but each one has its characteristic frequency peaks determined by the geometry of vocal tract during enunciation. The individual instances show dramatic random gyrations in the amplitude spectra. However, only the macroscopic spectral features are important. The use of Laguerre functions makes averages of the amplitude spectra. The recovered amplitude spectra show a substantial

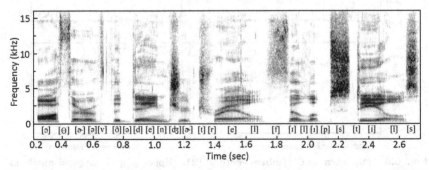

Fig. 6.4. Spectrogram of a sample sentence. Sentence a0001 in the ARCTIC databases, read by a male speaker bld.

134

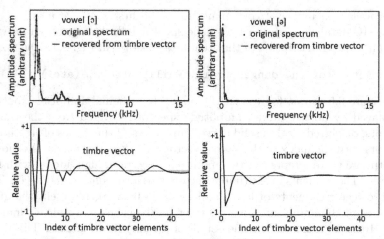

Fig. 6.5. **Spectra and timbre vectors (1).** Upper graphs: original amplitude spectra and the amplitude spectra recovered from timbre vectors. Lower graphs: the timbre vectors. Showing vowels [ɔ] and [ə].

smoothing. Nevertheless, all essential features are preserved.

The graphs of the timbre vectors of fricatives show a characteristic zigzag pattern. It is originated from the last large peak in each Laguerre function. As shown in Fig. 6.2, the sign of the last large peak is positive for Laguerre

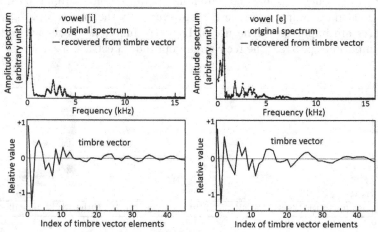

Fig. 6.6. **Spectra and timbre vectors (2).** Upper graphs: original amplitude spectra and the amplitude spectra recovered from timbre vectors. Lower graphs: the timbre vectors. Showing vowels [i] and [e].

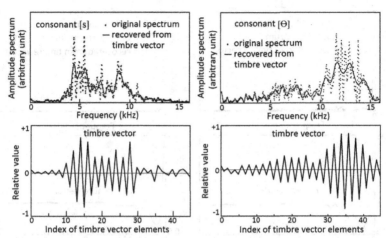

Fig. 6.7. Spectra and timbre vectors (3). Upper graphs: original amplitude spectra and the amplitude spectra recovered from timbre vectors. Lower graphs: the timbre vectors. Showing unvoiced consonants [s] and [θ].

functions of even order n, and negative for odd order n. Because the amplitude spectrum is always positive, in the zigzag region, the even elements are positive, and the odd elements are negative. The envelope of the zigzag features coincides with the overall features in the amplitude spectra.

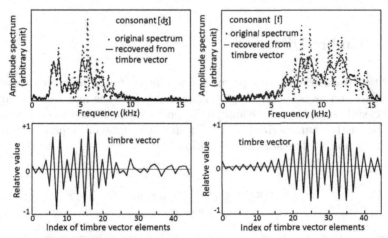

Fig. 6.8. Spectra and timbre vectors (4). Upper graphs: original amplitude spectra and the amplitude spectra recovered from timbre vectors. Lower graphs: the timbre vectors. Showing unvoiced consonants [dʒ] and [f].

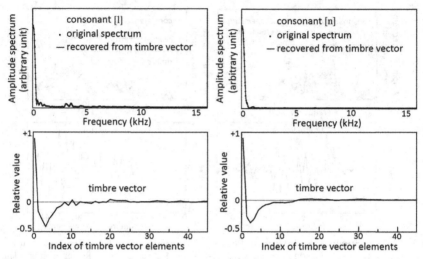

Fig. 6.9.   Spectra and timbre vectors (5). Upper graphs: original amplitude spectra and the amplitude spectra recovered from timbre vectors. Lower graphs: the timbre vectors. Showing voiced consonants [l] and [n].

The voiced consonants [l] and [n] are shown in Fig. 6.9. Both show some amplitudes at very low frequency. Not only the amplitude spectra, but also the timbre vectors are basically flat over the entire range of values. It is an evidence that for the four vowels, [ɔ], [ə], [i], and [e], the features at the high indices in the timbre vectors are non-trivial.

6.4   Timbre Distance

A critical parameter in any parametrical representation of speech signal is the distortion measure, or simply distance. Because the methods presented here cleanly separate timbre and pitch, the distortion measures defined here can be justly characterized as *timbre distance* between a pair of timbre vectors $\mathbf{x}$ and $\mathbf{y}$. The simplest definition is

$$d(\mathbf{x}, \mathbf{y}) = \sqrt{\sum_{n=0}^{n<N} (x_n - y_n)^2}, \qquad (6.17)$$

where x_n and y_n are the components of the two timbre vectors, or the arrays of the normalized Laguerre spectral coefficients $\mathbf{x}$ and $\mathbf{y}$. Because voiced frames and unvoiced frames are clearly separated, the timbre distance only applies to frames of identical type, which eliminates the comparison between apples and oranges by using the MFCC approach, see Section 6.1.1.

A possible alternative form, using a non-negative weight function $w(n)$, may improve the resolution power. The weight functions can be different for vowels and for unvoiced consonants:

$$d(\mathbf{x}, \mathbf{y}) = \sqrt{\sum_{n=0}^{n<N} w(n) \, (x_n - y_n)^2}. \tag{6.18}$$

It is straightforward to show that those definitions of distortion measures satisfy the three mathematical requirements in Section 6.1. Applications of timbre distances will be presented in Chapter 8.

Chapter 7

Waveform Recovery

In Chapter 6, we show that the amplitude spectrum of a timbron can be represented by a timbre vector, which contains a set of normalized Laguerre spectral coefficients, representing the instantaneous timber of the frame. Besides a timbre vector, the feature vector of a frame also contains the type of the frame (through the voicedness index), duration, and intensity, constituting a complete mathematical representation of a frame.

In this chapter, we will present the theory and procedure of how to recover the waveform of a timbron from a feature vector. By superposing the timbrons with the timing information of the glottal closure instants or segmentation points of choice – that is, the duration parameter in the feature vector – we can recover the entire speech signal. Because the timbre vector representation completely separates timbre and pitch, a large variety of voice transformation operations can be implemented, which will open a new way of doing speech synthesis and singing synthesis. A method for CD-quality speech coding algorithm can be developed. The pitch-synchronous spectral parameters create a new way of voice recognition, and a new approach to phonetics. We will present those technology applications in Chapter 8.

7.1 Phase Recovery

In Section 5.1.1, we show that because of the *causality property* of a timbron, the phase spectrum of a timbron is completely determined by its amplitude spectrum. In this section, we will present the details of how to recover the phase spectrum from the amplitude spectrum.

7.1.1 An analytic model

In this subsection, we present an analytic model of a piece of acoustic signal, a timbron triggered by a glottal closure at $t = 0$ with N formants. By studying that analytic model, we can gain a good understanding of the general properties of the amplitude spectrum and phase spectrum, and test the dispersion relations. The elementary acoustic wave is

$$x(t) = \sum_{n=0}^{n<N} C_n e^{-\beta_n t} \sin \alpha_n t \quad t \geq 0, \tag{7.1}$$

$$x(t) = 0, \qquad\qquad\qquad\qquad t < 0,$$

where C_n is the amplitude, β_n is the decay constant, and α_n is the central circular frequency of the n-th formant. The Fourier transform of the waveform $x(t)$ is

$$F(\omega) = \int_0^\infty x(t) e^{i\omega t} dt$$

$$= \frac{1}{2i} \sum_{n=0}^{n<N} C_n \left[\frac{1}{\beta_n - i(\omega + \alpha_n)} - \frac{1}{\beta_n - i(\omega - \alpha_n)} \right]. \tag{7.2}$$

The real part and imaginary part of $F(\omega)$ are

$$\mathrm{Re}\, F(\omega) = \sum_{n=0}^{n<N} \frac{C_n \alpha_n (\alpha_n^2 + \beta_n^2 - \omega^2)}{2 D_n}, \tag{7.3}$$

$$\mathrm{Im}\, F(\omega) = \sum_{n=0}^{n<N} \frac{C_n \alpha_n \beta_n \omega}{D_n}, \tag{7.4}$$

where the denominator D_n is

$$D_n = \left[\beta_n^2 + (\omega - \alpha_n)^2 \right] \left[\beta_n^2 + (\omega + \alpha_n)^2 \right]. \tag{7.5}$$

The amplitude spectrum $A(\omega)$ and the phase spectrum $\phi(\omega)$ are defined in a similar way as in Eqs. 7.28 and 7.29.

In the following, we should look at the general behavior of the amplitude spectrum and the phase spectrum. At low frequencies, when $\omega \ll \alpha_n$,

$$D_n = \left[\beta_n^2 + \alpha_n^2 \right]^2 + O(\omega^2). \tag{7.6}$$

The first term of amplitude $A(\omega)$ is determined by the first term of the real part $\mathrm{Re}\, F(\omega)$, with an error term $O(\omega^2)$,

$$A(0) = \frac{1}{2} \sum_{n=0}^{n<N} \frac{C_n \alpha_n}{\alpha_n^2 + \beta_n^2} + O(\omega^2). \tag{7.7}$$

For small ω, the imaginary part $\operatorname{Im} F(\omega)$ is

$$\sum_{n=0}^{n<N} \frac{C_n \alpha_n \beta_n}{[\beta_n^2 + \alpha_n^2]^2} \omega. \tag{7.8}$$

The first term of the phase is

$$\phi(\omega) \cong \frac{1}{A(0)} \sum_{n=0}^{n<N} \frac{C_n \alpha_n \beta_n}{[\beta_n^2 + \alpha_n^2]^2} \omega, \tag{7.9}$$

which is linear to the frequency ω.

At high frequencies, if $\omega \gg \alpha_n$ and $\omega \gg \beta_n$, then

$$D_n \cong \omega^4. \tag{7.10}$$

The real part $\operatorname{Re} F(\omega)$ is approximately

$$\operatorname{Re} F(\omega) \cong -\frac{1}{2\omega^2} \sum_{n=0}^{n<N} C_n \alpha_n, \tag{7.11}$$

and the asymptotic value of the imaginary part is

$$\operatorname{Im} F(\omega) \cong \frac{1}{2\omega^3} \sum_{n=0}^{n<N} C_n \alpha_n \beta_n. \tag{7.12}$$

Because the imaginary part decays much faster than the real part, the asymptotic value of the amplitude is the absolute value of the real part

$$A(\omega) \cong \frac{1}{2\omega^2} \sum_{n=0}^{n<N} C_n \alpha_n, \tag{7.13}$$

which has a characteristic inverse square relation to frequency ω.

Finally, the asymptotic values of the phase is

$$\phi(\omega) \cong \pi - \frac{K}{w}, \tag{7.14}$$

where the constant K is

$$K = \frac{2 \displaystyle\sum_{n=0}^{n<N} C_n \alpha_n \beta_n}{\displaystyle\sum_{n=0}^{n<N} C_n \alpha_n}. \tag{7.15}$$

Therefore, the general behavior of the phase spectrum at large frequencies is approaching π with an asymptote of the inverse of frequency.

7.1.2   Dispersion relation: an analytic example

In this subsection, we test the use of dispersion relations to recover the phase spectrum from the amplitude spectrum on the analytic model presented in the previous subsection.

The general formula for computing the phase spectrum from the amplitude spectrum using dispersion relations is Eq. 5.4. By dividing both the numerator and the denominator by 2π, we have the following expression in terms of frequency,

$$\phi(f) = -\frac{1}{\pi} \lim_{\epsilon \to 0} \left[\int_{-\infty}^{f-\epsilon} \frac{\ln A(f')}{f'-f} df' + \int_{f+\epsilon}^{\infty} \frac{\ln A(f')}{f'-f} df' \right]. \tag{7.16}$$

Practically, the frequency is digitized with a frequency interval δ, and the integration becomes a sum. Denote the logarithm of amplitude as

$$\Lambda(f) = \ln A(f), \tag{7.17}$$

and then Eq. 7.16 becomes

$$\phi(f) = -\frac{1}{\pi} \sum_{k=1}^{\infty} \frac{\Lambda(f+k\delta) - \Lambda(f-k\delta)}{k}. \tag{7.18}$$

The sum is evaluated for both positive frequencies and negative frequencies. The amplitude spectrum is an even function on the entire axis of frequency. Figure 7.1 shows a typical amplitude spectrum of a vowel. According to Eq. 7.13, at higher frequencies, typically for $f \gg 5$ kHz, the amplitude spectrum falls off as the inverse square of frequency.

Notice that the logarithm falls off very slowly. To achieve a high accuracy, a large number of terms are required, making computation quite slow. In the following, we implement three approximations which would result in sufficient accuracy but requires much less computation time.

First, it is not necessary to compute the phase spectrum over the entire range of frequencies. Because for voiced sounds, the amplitude spectrum above 5 kHz is weak. Therefore, phase is only important for frequency lower than 5 kHz.

Second, for voiced frames at higher frequencies, for example greater than 10 kHz, the amplitude spectrum is weak and falls off as the inverse square of frequency. Therefore, after a cutoff frequency f_0, or an index k_0, the amplitude spectrum can be approximated by

$$A(k) \approx A(k_0) \frac{k_0^2}{k^2} \qquad k > k_0. \tag{7.19}$$

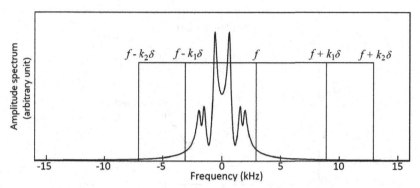

Fig. 7.1. Amplitude spectrum on the entire frequency axis. Right side: dotted curve: original; thin curve: recovered. Left side: timbrons.

As shown in Fig. 7.1, for $k \gg k_1$, in the positive frequency range, the amplitude spectrum falls into the inverse square range.

Third, the integration need not to go indefinitely. For $|k| > k_2$, for both positive frequency and negative frequency, the amplitude spectrum follows an inverse-square relation, the sum can be evaluated analytically. In fact, by using a cutoff number K, or a cutoff frequency $f_K = K\delta$, the sum in Eq. 7.18 can be split into two parts,

$$\phi(f) = -\frac{1}{\pi} \left[\sum_{k=1}^{K} \frac{\Lambda(f + k\delta) - \Lambda(f - k\delta)}{k} + R_K \right], \qquad (7.20)$$

where the residual term R_K is

$$R_K = \sum_{k=K}^{\infty} \frac{\Lambda(f + k\delta) - \Lambda(f - k\delta)}{k}. \qquad (7.21)$$

Because the amplitude spectrum falls off as f^{-2},

$$\Lambda(f + k\delta) = \text{const.} - 2\ln k\delta. \qquad (7.22)$$

We have

$$R_K = \sum_{k=K}^{\infty} \frac{4f}{k^2 \delta} \approx \frac{4f}{K\delta} = \frac{4f}{f_K}. \qquad (7.23)$$

Coming back to Eq 7.20, we have

$$\phi(f) = -\frac{1}{\pi} \left[\sum_{k=1}^{K} \frac{\Lambda(f + k\delta) - \Lambda(f - k\delta)}{k} + \frac{4f}{f_K} \right]. \qquad (7.24)$$

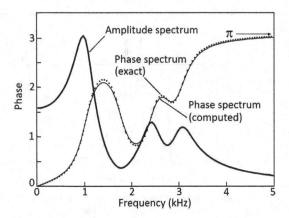

Fig. 7.2. Phase recovered from dispersion relations. Right side: dotted curve: original; thin curve: recovered. Left side: timbrons.

Figure 7.2 shows an analytic example: the amplitude spectrum, the exact phase spectrum, and the phase spectrum computed from the amplitude spectrum using dispersion relations with the approximations described above. As shown, the computed phase spectrum using dispersion relations agrees well with the exact phase spectrum of the analytic model.

7.1.3 Phase spectrum from speech signals

According to the Fourier theorem, by expanding a discrete periodic function with N points in each period into a complex Fourier series, the original function can be exactly reproduced from the Fourier coefficients. Let the periodic function be $X(n)$, where $0 \leq n \leq N$, and $X(0) = X(N)$, the Fourier coefficients are

$$F(k) = \frac{1}{N} \sum_{n=0}^{N-1} X(n) \exp\left\{-\frac{2\pi i k n}{N}\right\}. \tag{7.25}$$

And the function $X(n)$ can be recovered *exactly* by

$$X(n) = \sum_{k=0}^{N-1} F(k) \exp\left\{\frac{2\pi i k n}{N}\right\}. \tag{7.26}$$

The complex Fourier coefficients $F(k)$ can be written in terms of an amplitude spectrum and a phase spectrum. Both are real.

$$F(k) = A(k)\, e^{i\phi(k)}, \tag{7.27}$$

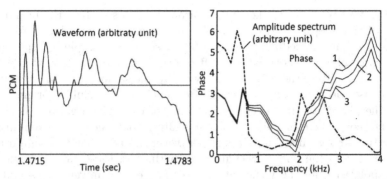

Fig. 7.3. Acoustic wave, amplitude spectrum and phase spectrum. By shifting the origin of the acoustic wave in a pitch period by one or two single sampling points, the phase spectrum shifts linearly to frequency and the number of the sampling-point shift.

where the amplitude is

$$A(k) = \left[(\operatorname{Re} F(k))^2 + (\operatorname{Im} F(k))^2 \right]^{\frac{1}{2}}, \tag{7.28}$$

and the phase is

$$\phi(k) = \arctan(\operatorname{Im} F(k), \operatorname{Re} F(k)), \tag{7.29}$$

where during computation, the function atan2 in ANSI C is used, which has two arguments.

While the amplitude $A(k)$ is independent of the starting point of a periodic function, the phase $\phi(k)$ does depend on the starting point of a periodic function, see Section 5.1.3. By shifting the origin by m, the new Fourier coefficient, denoted by $F_m(k)$, is

$$
\begin{aligned}
F_m(k) &= \frac{1}{N} \sum_{n=0}^{N-1} X(n+m) \exp\left\{ -\frac{2\pi i k n}{N} \right\} \\
&= \frac{1}{N} \sum_{n=0}^{N-1} X(n) \exp\left\{ -\frac{2\pi i k (n-m)}{N} \right\} \\
&= \exp\left\{ \frac{2\pi i k m}{N} \right\} F(k).
\end{aligned}
\tag{7.30}
$$

Comparing with Eq. 7.29, with a shift of origin by m, the phase shifts by

$$\phi(k) \longrightarrow \phi(k) + \frac{2\pi k m}{N}. \tag{7.31}$$

In other words, *the phase shift is a linear function of k and m.* In the following, we consider an example, a pitch period in ARCIC database, sentence a0002, read by male speaker bdl, at around 1.475 sec. It is the beginning part of vowel [e] in word "case". Figure 7.3 (A) is the waveform after the ends-matching procedure. Figure 7.3 (B) shows the amplitude spectrum and three phase spectra with a shift of origin, marked by 1, 2, and 3. As shown, by shifting the origin of the wave by one sampling point each time, the phase spectrum shift is linear to frequency and the shift of origin.

The phase at zero frequency is worthy of close investigation. Because the DC component in the voice signal is filtered out, the value of amplitude spectrum at zero frequency is zero. However, because both the amplitude spectrum and the phase spectrum are continuous, those values should be extrapolated from the values at low frequencies. A universal observation is that the extrapolated value of phase at zero frequency is always a finite value, roughly equals π. We will return to this point in Section 7.1.4.

7.1.4   Dispersion relation: voice data

In this section, we show the phase spectrum of real voice signals can be recovered with reasonable accuracy using dispersion relations. Figure 7.4 shows an example, vowel [ɔ] in word "Tom", spoken by a male speaker bld, in ARCTIC databases, sentence a0002. Figure 7.4 (A) is the original amplitude spectrum and phase spectrum. The phase at zero frequency is extrapolated from phase values of low frequencies. The value is close to π. Figure 7.4 (B) is recovered spectra. The amplitude spectrum is a smoothed version of the original, and all basic features are well reproduced. The phase spectrum recovered from amplitude spectrum using dispersion

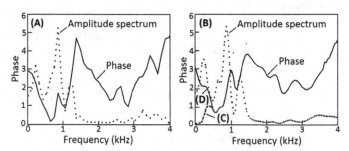

Fig. 7.4.   Phase spectrum recovered from dispersion relations. (A), original experimental data. At very low frequency, the phase approaches π. (B), phase spectrum calculated from amplitude spectrum using dispersion relations. At low frequencies, especially when $f < 0.5$ kHz, the frequency calculated using dispersion relations approaches 0, as expected from the analytic properties of amplitude spectrum and phase spectrum, shown by dashed curve (C). A phase correction term for the d'Alembert wave of the glottal flow is needed to obtain a phase spectrum close to the original phase spectrum at low frequencies, (D). See Section 7.1.5.

relations also reproduces all basic features. Above 4 kHz, the amplitude spectrum is weak; and the phase spectrum becomes noisy, which is not shown. At zero frequency, the phase spectrum recovered from dispersion relations approaches zero, as shown by the thick dashed curve (C). It is expected that since the amplitude spectrum is an even function of frequency, the phase spectrum should be an odd function of frequency. Clearly, it deviates from the observed phase spectrum, which approaches about π near the zero-frequency point. The discrepancy is related to the d'Alembert wave of the glottal flow, which can be corrected by adding a correction term in low-frequency region to become curve (D), as shown in the following section.

7.1.5   d'Alembert wave of glottal airflow

The difference near $f = 0$ between the measured phase and the phase recovered from amplitude spectrum using Kramers-Knonig relations, as shown in Fig. 7.4, originates from the d'Alembert wave of the glottal airflow, as shown in Fig. 7.5. Figure 7.5 (A) shows the airflow when the glottis is open. Outside the mouth, air flows outwards. (B) shows the airflow after the glottal closure but before the d'Alembert wavefront of the glottal closure reaches the mouth. An outward airflow outside the mouth continues. (C) shows the air flow immediately after the d'Alembert wavefront of the glottal closure reaches the mouth. Inside the vocal tract, air is rarefied and the perturbation pressure is negative. The air outside the mouth rushes in to make a sharp pulse of airflow into the vocal tract.

By looking into the real phase spectrum in Fig. 7.4, one finds that there

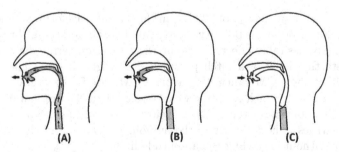

Fig. 7.5. d'Alembert wave of glottal airflow. (A) when the glottis is open, airflow immediately outside the mouth is outward. (B) Immediately after a glottal closure, before the zero-velocity d'Alembert wavefront reaches the mouth, the outward airflow continues. (C) immediately after the zero-velocity d'Alembert wavefront reaches the mouth, a strong inward airflow starts, which is the sharp peak of the voice signal at the starting point of a pitch period.

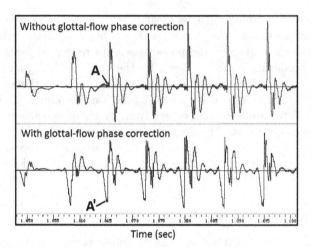

Fig. 7.6. Glottal-flow phase correction. Upper half: Using the phase spectrum recovered by the Kramers-Kronig relations, the starting feature of a pitch period is very sharp, see point A. Lower half, after adding a phase correction term, Eq. 7.32, a slowly varying feature in the opposite polarity is generated, representing the d'Alembert wave of the glottal flow function, which also softens the sharp features at the beginning of a pitch period.

is a phase difference in the low-frequency range, with a maximum value of π near $f = 0$. Experiments showed that by adding a correction term to the phase spectrum as shown below, the problem is resolved:

$$\Delta\phi = \pi\left(1 - \frac{f}{f_0}\right), \qquad f < f_0. \qquad (7.32)$$

With a cutoff frequency f_0 of 500 Hz, in the output waveform, an additional broad peak of opposite polarity is generated. The additional broad peak has a highly desirable effect. Without the correction term, the amplitude of the starting peak of a pitch period is very high. It often goes beyond the limit of the PCM range. To avoid PCM cutoff, the overall intensity must be reduced, which results a weaker voice than the original one. After adding this correction term, the waveform become more symmetric over both polarities. The intensity of the outgoing voice can match the original intensity without exceeding the amplitude limit.

Figure 7.6 shows the recovered waveform, without the glottal-flow phase correction, upper part; and with the glottal-flow phase correction, lower part. As shown, without a glottal-flow phase correction, at the starting point of a pitch period, A, there is a sharp pulse, and often with a very large peak. With glottal-flow phase correction, A', a smooth feature before the sharp peak is generated, which is the d'Alembert wave of the glottal flow. The sharp peak is reduced. The waveform is more symmetric.

7.2 Timbron Recovery

In this Section, we present a number of examples of vowels, to show how individual timbrons are recovered from timbre vectors, see Figs. 7.7 and 7.8. All examples are taken from the ARCTIC databases.

Figure 7.7 shows three back vowels, [ɑ], [u], and [o]. On the left-hand side, there are original amplitude spectra of the vowels, dotted curves; and the amplitude spectra recovered from the timbre vectors, solid curves. As

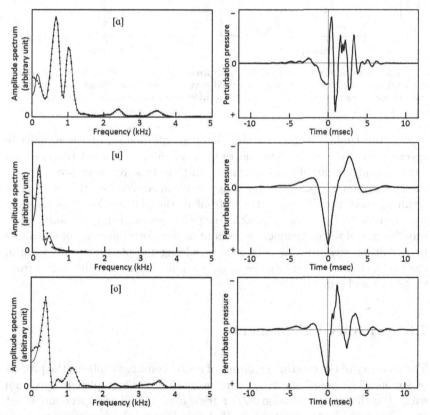

Fig. 7.7. Timbrons recovered from timbre vectors: back vowels. Right side: dots: original amplitude spectrum; thin solid curve: recovered amplitude spectrum. Left side: timbrons recovered from the amplitude spectrum and the phase spectrum.

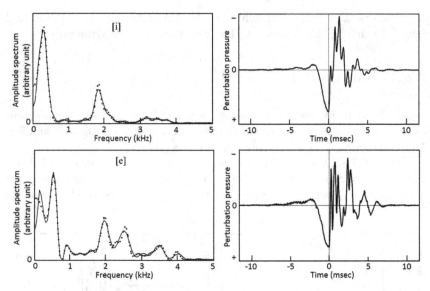

Fig. 7.8. Timbrons recovered from timbre vectors: front vowels. Right side: dots: original amplitude spectrum; thin soild curve: recovered amplitude spectrum. Left side: timbrons recovered from the amplitude spectrum and the phase spectrum.

shown, the timbre vector contains detailed information of the amplitude spectrum of the vowels. On the right-hand side, individual timbrons recovered from the amplitude spectra, with the phase recovery procedure in the last section, are shown. Following the usual convention, the sharp peak resulting from the d'Alembert wavefront of the glottal closure is pointing up. It is actually a sharp peak of negative perturbation pressure. The broad peaks of the d'Alembert wavefront of the glottal flow are of a positive perturbation pressure. As shown, after adding the phase correction term, the peaks on both polarities are roughly equal. Figure 7.8 shows two front vowels, [i] and [e].

7.3   Unvoiced Consonents

The recovery of the starting timbrons of a stop consonant follows the process of vowels. The phase spectrum is recovered from the amplitude spectrum using the dispersion relations. For fricatives, the phase spectrum is set to be *random*. Experiments show that this is the only procedure required to recover a fricative. In contrast to the reproduction procedure based on the source-filter model, no noise source is required. The simplicity of this process enables the definition of *fractional voicedness index*, to be used for voiced fricatives, see Subsection 7.4.2.

7.4 Recovery of Entire Voice Signal

The entire voice signal can be recovered following the superposition principle, by adding each timbron to the output voice signal consecutively according to the timing information provided by the timbre vectors.

Figure 7.9 shows an example of the spectrograms of an original speech signal and a recovered speech signal. The original sentence, a0005 of ARCTIC databases, spoken by a female speaker slt. As shown, the spectrogram is reproduced in great detail. There are subtle differences. However, those minor differences would not make a difference in listening.

Figure 7.10 shows a part of the original waveform and the recovered waveform. As shown, because the waveform is recovered period by period, the waveform of each pitch period is individually recovered. Especially, the d'Alembert waveform of the glottal flow, the broad peak below the zero line before the sharp peak P of the d'Alembert wavefront of the glottal closing, marked as G, is also reproduced for each pitch period.

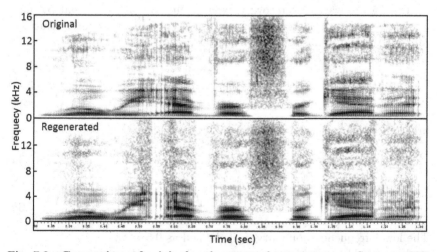

Fig. 7.9. Comparison of original and recovered spectrograms. Sentence a0005 of the ARCTIC databases, speaker slt: "Will we ever forget it."

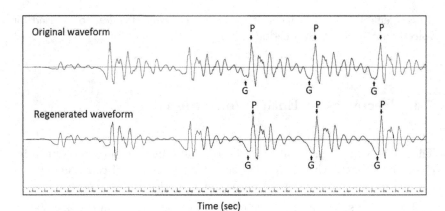

Fig. 7.10. **Comparison of original and recovered waveforms.** G: the glottal flow waveform. P: the starting pulse of the d'Alembert wavefront of a glottal closing.

7.4.1   Breathiness index

In the preceding sections, we show that the phase spectrum of voiced sounds can be regenerated from the amplitude spectrum using dispersion relations, also called Kramers-Kronig relations. For unvoiced fricatives, the phase spectrum is random. However, the power of the voiced sounds is concentrated in the low-frequency range, typically 0 to 4 kHz. In the frequency range higher than 4 kHz, the phase spectra of the voiced periods are rather random in the original signals, but perception is not affected.

During the waveform recovery experiments, it was found that by using a mixture of determinstic phase spectrum (from Kramers-Kronig relations) ϕ_{KK} and random phase spectrum ϕ_{RD}, for example,

$$\phi(f) = \begin{cases} \phi_{\mathrm{RD}} & : \quad f > f_2 \\ (1 - p(f))\phi_{\mathrm{RD}} + p(f)\phi_{\mathrm{KK}} & : \quad f_1 < f < f_2 \ , \\ \phi_{\mathrm{KK}} & : \quad f < f_1 \end{cases} \qquad (7.33)$$

breathiness effect can be generated. The function $p(f)$ in Eq. 7.33 is

$$p(f) = 3x^3 - 2x^2, \qquad (7.34)$$

with

$$x = \frac{f - f_1}{f_2 - f_1}. \qquad (7.35)$$

The lower frequency boundary f_1 is defined as $0.5 \times f_2$.

Figure 7.11 shows the spectrogram of the recovered voice signals with a different upper frequency bound f_2. In Fig. 7.11 (A), the upper bound of a

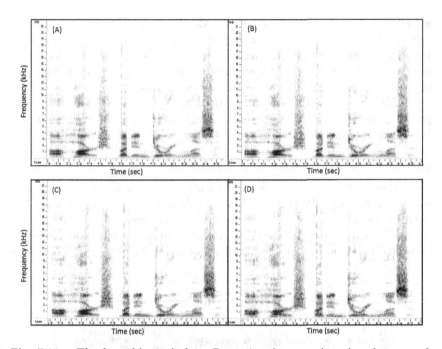

Fig. 7.11.   The breathiness index. By setting the upper boundary frequency f_2 for the phase spectrum determined by the Kramers-Kronig relations, variable effects of breathiness can be generated, up to the extreme case of whispering. (A), $f_2 = 20$ kHz. (B), $f_2 = 5$ kHz. (C), $f_2 = 2$ kHz, a strong breathy character. (D), $f_2 = 0$ kHz, whispering.

deterministic phase spectrum f_2 was set to be 20 kHz. The periodicity of the voiced signals can be clearly observed in the high-frequency region. In Fig. 7.11 (B), f_2 was set to be 5 kHz. The periodicity of the voiced signals in the high-frequency region is lost. However, informal listening does not detect any audible differences. By lowering f_2 to 2 kHz, as shown in Fig. 7.11 (B), a pronounced breathy character is heard. With a zero f_2, the entire speech is set to have random phase spectra. All pitch period boundaries are lost. A whispering voice is generated.

To summarize, by setting the upper bound frequency f_2 for the deterministic phase spectrum, effects of breathiness can be generated, up to the extreme case of whispering: a purely breathy voice without pitch periods.

7.4.2   Fractional voicedness index

By setting a global upper boundary frequency f_2 for the phase spectrum determined by the Kramers-Kronig relations, various degrees of breathiness for the entire sentence can be generated. The upper bound frequency f_2 can also be set for individual pitch periods to generate the effect of a *fractional voicedness index*. For the following two cases in speech signals, the implementation of fractional voicedness index can improve the naturalness of reproduced speech: The voiced fricatives such as [z], [ʒ], [ʋ] and [ð], and the transitional periods between a vowel and an unvoiced fricative. To process large amount of speech data, the fractional voicedness index should be set up automatically. In the following, we show that within the framework of the timbre vector formalism, a well-defined procedure of setting a fractional voicedness index can be implemented.

Figure 7.12 shows a case of a voiced fricative [z] in a word *magazine*, in sentence a0088 of ARCTIC Databases, spoken by a female speaker slt. (A) is the voice signal. The voiced fricative [z] is between 2.7 sec and 2.8 sec. The periodicity is clearly observed, and the automatic detection procedure presented in Section 5.4 identifies the pitch marks. Therefore, initially, those

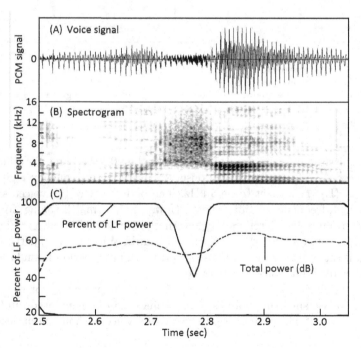

Fig. 7.12. Detection of voiced fricatives. (A) the voice signal, sentence a0088 of ARCTIC Databases, speaker slt. From 2.7 sec to 2.8 sec is a voiced fricative [z]. (B) the spectrogram, where the phase spectrum of the voiced fricative is random above 2 kHz. (C) is the total energy and the percentage of low frequency energy in each pitch period.

periods are marked as voiced. However, the spectrogram of the voiced fricatives looks very different from the vowels. For the vowels, the pitch periods are clearly identifiable. For voiced fricatives, the pitch-period boundaries are blurred, except for the very low frequency components, typically below 2 kHz. Therefore, *the phase spectrum of the voice signal is random above 2 kHz*. On the other hand, for voiced fricatives, the power in the high-frequency region, typically higher than 4 kHz, is strong; whereas the power in the low-frequency region, typically lower than 4 kHz, is weak. Because the segmentation process in Section 5.4 already segments the entire speech signal train into pitch periods or its equivalence in the unvoiced sections, for each frame, the total energy and the energy in the high-frequency range and the low-frequency range can be separated. Specifically, for each pitch period, the following sums of the amplitude spectra are computed,

$$\epsilon_{\text{L}} = \sum_{f \leq f_0} A(f)^2, \tag{7.36}$$

and

$$\epsilon_{\text{H}} = \sum_{f > f_0} A(f)^2, \tag{7.37}$$

where f_0 can be set as for example 4 kHz. Then, two quantities are derived: the total energy in a logarithmic scale

$$W = 4.343 \ln(\epsilon_{\text{H}} + \epsilon_{\text{L}}), \tag{7.38}$$

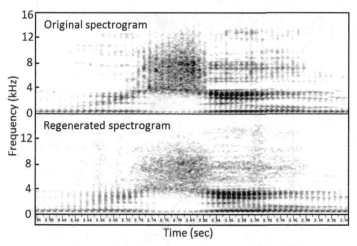

Fig. 7.13. Implementation of the recovery of voiced fricatives. (A) the original spectrogram, sentence a0088 of ARCTIC Databases, speaker slt. (B) the spectrogram of the regenerated speech.

and the percentage of low-frequency energy

$$\rho = 100 \, \frac{\epsilon_L}{\epsilon_H + \epsilon_L}. \tag{7.39}$$

Those quantities are displayed on Figure 7.12(C). Note that the unit of the total energy W is in decibels up to an overall constant. As shown in Fig. 7.12, for vowels, the power is concentrated in the low-frequency range, typically $f < 4$ kHz. For voiced consonants, a substantial portion of voice energy is in the high-frequency range. Therefore, a high percentage of low-frequency power is a signature of full voicedness. If the percentage of low-frequency power is much less than 100, the pitch periods are characterized by a fractional voicedness.

During voice reproduction, a fractional voicedness index is implemented as an upper boundary frequency for the Kramers-Kronig phase spectrum lower than 5 kHz. Figure 7.13 shows the original spectrogram and the reproduced spectrogram of a section of speech signals in Fig. 7.12. As shown in Fig. 7.13, the spectrogram of the voiced fricative [z] is made similar to that of the original spectrogram.

Next, we consider the case of a transition between a vowel and an unvoiced fricative, a section in sentence a0088 of the ARCTIC databases, spoken by a male speaker bdl, see Fig. 7.14. Between 1.88 sec and 1.92 sec, the pure unvoiced consonant [s] occurs. Between the unvoiced consonant and

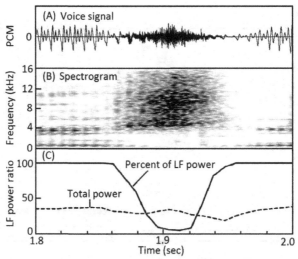

Fig. 7.14. Transitional fractional voicedness index. (A), the original spectrogram, sentence a0088 of ARCTIC Databases, speaker bdl. (B), the spectrogram of the regenerated speech. (C), the percentage of low-frequency power.

the vowels, there are two pitch periods in 1.86 sec and 1.88 sec as well as two pitch periods between 1.92 sec and 1.94 sec which are clearly voiced fricatives, from the voice signal and from the spectrogram. On the other hand, for the unvoiced fricatives, the percentage of low-frequency power is nearly zero. And for the transitional pitch periods, the percentage of low-frequency power is between 10% and 90%. Again, the percentages of low-frequency power is a good sign of fractional voicedness; and during voice regeneration, using a low upper boundary of Kramers-Kronig phase spectrum, the effect of the fractional voicedness can be implemented.

The introduction of fractional voicedness index and its implementation through a lower upper bound of Kramers-Kronig phase spectrum can improve the reproduction of voiced fricatives and the transition between vowels and unvoiced fricatives. However, it is not necessarily to make it too complicated. In the case of speech coding, as we will show in Section 8.2, a general identification of fractional voicedness is already better than not to have it. In other words, one can assign a two-bit type identification as follows:

silence	00
unvoiced fricative	01
voiced fricative	10
voiced and stop	11

Furthermore, using the timbre fusing procedure, as shown in Section 12.1, when a vowel and an unvoiced fricative are connected, fractional voicedness indices can be automatically generated at the transitional periods.

7.4.3　Jitter and shimmer

Because in the timbre vector parameterization, for each pitch period, pitch period and intensity are components of a feature vector, see Section 6.2.3; jitter and shimmer can be easily implemented.

First, if the original jitter or shimmer is too large, one of them or both can be reduced by taking a running average over all voiced sections in a feature vector file. If the pitch periods (timing parameter τ) in a voiced section are $\tau_0, \tau_1, \tau_n, \ldots \tau_N$, then let the new pitch periods $\overline{\tau_n}$ be

$$
\overline{\tau_n} = \begin{cases} \tau_0 & : \quad n = 0 \\ \dfrac{\tau_{n-1}}{4} + \dfrac{\tau_n}{2} + \dfrac{\tau_{n+1}}{4} & : \quad 0 < n < N \\ \tau_N & : \quad n = N \end{cases} \tag{7.40}
$$

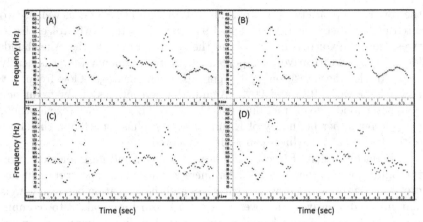

Fig. 7.15. Jitter: random fluctuation of pitch periods. The pitch data are obtained from Wavesurfer. (A) original pitch. (B) after jitter reduction by doing a running average. (C) after adding a 3% jitter. (D) after adding a 7% jitter.

the jitter of the voice is reduced. Figure 7.15(A) is the original pitch. (B) gives the pitch values after running averaging.

To increase jitter, a random term is added. The new pitch period for the period n becomes

$$\overline{\tau_n} = \tau_n(1 + \delta R_n), \qquad (7.41)$$

where δ is the amount of jitter, and R_n is a random number between -1 and $+1$. Figure 7.15(C) shows the pitch values after adding a 3% jitter, and (D) shows the pitch values after adding a 7% jitter.

Similarly, by making a running average over the amplitude in a series of feature vectors, the shimmer of a sentence is reduced. By adding a random term to the amplitude, shimmer is increased.

Chapter 8

Applications

Much of the speech and voice technology depends on some type of parameterization of speech and voice. Non-parametric methods, such as the pitch-synchronous overlap add technique (PSOLA) [9, 65, 66], which works in time domain, can only make very limited modifications to the natural speech, with applications confined in, for example, unit-selection TTS systems. To date, the most widely used parameterization methods are linear predictive coding (LPC) [58] and mel-frequency cepstral coefficients (MFCC) [24]. It is known that the vocoders based on LPC and MFCC can only produce reconstructed speech of rather poor quality, for example, see Chapter D24 of *Springer Handbook of Speech Processing* [6].

The pitch-synchronous parameterization method based on timbre vectors presented in Chapters 5, 6, and 7 is a replica of human voice production process. As shown in Part I, for voiced sounds, speech signal is generated pitch-period by pitch-period, and each pitch period starts with a glottal closing event. The duration of a pitch period, defined as the time interval between two consecutive glottal closing moments, varies constantly.

By utilizing the pitch-synchronous parameterization method, speech and voice technology could be significantly improved. As a starting point, five U.S patents were issued in 2014 and 2015, disclosing systems and methods of its applications in voice transformation [12], speech coding [13], speech recognition [14], speech synthesis [11], and prosody generation [10].

8.1 Voice Transformation

Voice transformation refers to the various modifications one may apply to the sound produced by a person, speaking or singing [90, 91]. However, until recently, highly versatile and highly natural voice transformation has not been achieved. Stylianou [90] attributed the shortcomings to the lack of understanding of the physics of speech production:

> Voice transformation involves signal processing and the physics
> (or at least the understanding) of the speech production process

and natural language processing. Driven mainly by its applications, signal processing has evolved faster than the physics
of speech processing, even giving the impression that signal
processing alone may be required to achieve high-quality voice
transformation. To an external observer, this is similar to the
problem of how to make an omelet without eggs.

As a physicist working on speech technology for years, I agree with this
statement. The legacy signal processing methods, LPC and MFCC, do not
reflect the physics of human voice production. By designing a parameterization method based on an accurate understanding of the physics of voice
production, better voice transformation algorithms can be designed.

In Part I of this book, we presented the eggs, the *physics* of human
voice production. And in Part II, the cooking wares are provided. Figure 8.1 shows a recipe of how to make an omelet *with* eggs: a method
for voice transformation [12]. As shown, the input voice signal, in PCM
form, is first multiplied by an asymmetric window, see Fig. 5.6, Eqs. 5.22,
and 5.23, to generate a profile function. The peaks of the profile function
are taken as the segmentation points for voiced sections. The segmentation points are extended into unvoiced sections to form a complete list of
pitch-synchronous segmentation points. The voice signal is then segmented
into pitch-synchronous frames. The two ends of each frame are equalized

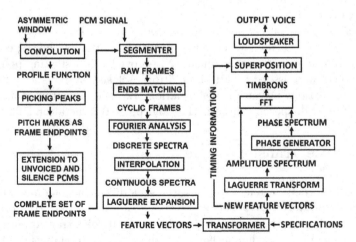

Fig. 8.1. Schematics of a voice transformation system. The input voice signal is
first segmented into pitch-synchronous frames. Then the ends of each frame are equalized
to make it cyclic. A Fourier analysis unit and a Laguerre function unit convert the signal
of each pitch-synchronous frame into a feature vector. By transforming the set of feature
vectors according to the specifications, a transformed voice signal is recovered [12].

with an ends-matching procedure to make it cyclic, see Section 5.2. Fourier analysis is then applied to those cyclic frames. Initially, the dimension of the amplitude spectrum is one half of the dimension of the frame. Using a truncated Whittaker-Shannon interpolation method, the dimension of the amplitude spectrum is expanded to a fixed number, typically 512. Using Laguerre functions, which have a weight function concentrated in the frequency range 0.3 kHz to 5 kHz, a convenient set of spectral coefficients is generated. The normalized Laguerre spectral coefficients, consist of a timbre vector, together with the voicedness index, the duration of the frame, and the rms amplitude, form a *feature vector* of the frame.

The above process literally decomposes the voice into feature vectors, similar to decomposing a piece of material into atoms. The set of feature vectors can be converted back to the original voice signals by first recovering the amplitude spectrum using Laguerre functions, and then recovering the phase spectrum using Kramers-Kronig relations. A FFT unit then converts the pair of amplitude spectrum and phase spectrum into an elementary acoustic wave, or a timbron. According to the timing information in the feature vector, those timbrons are superposed to recover the original voice. However, the set of feature vectors can be rearranged according to a set of specifications to become a new set of feature vectors. And the voice recovered from the new feature vectors can have a variety of voice transformations.

In the following subsections, we present three examples of voice transformation: pitch contour modification, change of speaker identity, and change of speaking speed.

8.1.1 Pitch contour modification

As a first example, we show how to change the pitch contour of a recorded sentence according to a desirable function of time to generate a new speech signal. First, the sentence is segmented into N frames, the n-th frame is characterized by a set of parameters; v_n, the voicedness index; τ_n, the pitch period; α_n, the amplitude; and c_n, the timbre vector; where $0 \leq n < N$. Using the procedure in Chapter 7, the original voice can be regenerated from that set of parameters.

The starting time of the 0-th frame is $t_0 = 0$. The starting time of the n-th frame is

$$t_n = \sum_{j=0}^{j<n} \tau_j, \tag{8.1}$$

and the total time of the sentence is

$$t_N = \sum_{j=0}^{j<N} \tau_j. \tag{8.2}$$

Suppose we want to change the pitch contour to a function of time over the entire duration of the sentence,

$$f(\tilde{t}), \quad 0 < \tilde{t} < t_N, \tag{8.3}$$

that is, to generate a new set of parameters for each new frame, $\tilde{v}_m$, $\tilde{\tau}_m$, $\tilde{\alpha}_m$ and $\tilde{c}_m$, where $0 \leq m < M$. In general, the number of frames M is not the original number N, but the total time of the sentence equals the original time,

$$\sum_{m=0}^{m<M} \tilde{\tau}_m = t_N. \tag{8.4}$$

For the first frame, the parameters are

$$\begin{aligned}
\tilde{v}_0 &= v_0 \\
\tilde{\tau}_0 &= \frac{1}{f(0)} \\
\tilde{\alpha}_0 &= \alpha_0 \\
\tilde{c}_0 &= c_0.
\end{aligned} \tag{8.5}$$

At the end of the first frame, the time is moved forward to $\tilde{t}_1 = \tilde{\tau}_0$. The frequency of the next frame is $f(\tilde{t}_1)$. It is not necessarily at a time point of the original frames. Therefore, the parameters have to be interpolated. In general, if the end of the new m-th frame is $\tilde{t}_m$, which is defined as

$$\tilde{t}_m = \sum_{j=0}^{j<m} \tilde{\tau}_j, \tag{8.6}$$

and it is between the n-th ending point and the $(n+1)$-th ending point of the original record,

$$t_n < \tilde{t}_m < t_{n+1}, \tag{8.7}$$

using a interpolation parameter δ

$$\delta = \frac{\tilde{t}_m - t_n}{\tau_n}, \tag{8.8}$$

the parameters of the $(m + 1)$-th new frame are

$$\tilde{v}_{m+1} = (1 - \delta)v_n + \delta v_{n+1}$$
$$\tilde{\tau}_{m+1} = \frac{1}{f(\tilde{t}_m)}$$
$$\tilde{\alpha}_{m+1} = (1 - \delta)\alpha_n + \delta\alpha_{n+1}$$
$$\tilde{\mathbf{c}}_{m+1} = (1 - \delta)\mathbf{c}_n + \delta\mathbf{c}_{n+1}.$$

$$(8.9)$$

The above algorithm can be easily programmed on a computer. An important note is as follows: In the timbre vector format, the voiced frames and the unvoiced frames are treated equally. The only difference during regeneration is in the upper limit of the deterministic phase spectrum. Furthermore, even if in the original timbre vector records, all voicedness indices are binary, fractional voicedness indices can be generated in the interpolation process. The result is a smoother transition between voiced and unvoiced sections.

In the following, we will demonstrate how to change the speaking mode of a sentence, a0207 in ARCTIC databases, "How much was it", spoken by a female speaker slt. The original speech is rather flat, sounds like a statement rather than a question. See Fig. 8.2(A) for the pitch contour as displayed by Wavesurfer. Using the algorithm presented above, the mode

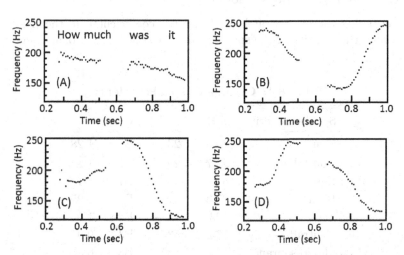

Fig. 8.2. Voice transformation by changing pitch contours. (A) the original speech, sentence a0207, "How much was it", spoken by female speaker slt. Although the text is a question, the pitch contour is rather flat, and sounds like a statement. (B) transformed into a strong question, "How much is IT?" (C) another mode of question, "How much WAS it?" (D) a further mode of question "How MUCH was it?"

of the sentence can be changed.

First, by significantly raising the pitch and intensity of the last word, the sentence is changed to a clear question, see Fig. 8.2(B),

> How much was IT?

Next, by raising the pitch and the intensity of the word "was", the focus of the question is placed on the action in the past, see Fig. 8.2(C),

> How much WAS it?

Such a question is asked in the case of selling the item, and to estimate the profit or loss. Finally by raising the pitch and stress of the word "much", the focus of the question is again shifted to the amount of money paid, see Fig. 8.2(D),

> How MUCH was it?

This implies a question regarding to the fairness of the amount being paid, whether it is overpaid or underpaid.

Those speech samples are posted on www.columbia.edu/~jcc2161/ under Human Voice.

8.1.2   Change of speaker identity

By changing the overall pitch level and the coefficient κ in the Laguerre function, the identity of the speaker can be changed. As a demo, sentence

Table 8.1: Change of speaker identity

Speaker identity	$\bar{f}$(Hz)	MIDI	note	κ
Giant	30.8	23	B0	1.28
Contrabass	43.7	29	F1	1.2
Bass	61.5	35	B1	1.1
Baritone	87.3	41	F2	1.05
Tenor	123	47	B2	1.0
Contralto	175	53	F3	0.9
Mezzo-soprano	247	59	B3	0.8
Soprano	349	65	F4	0.72
Child	494	71	B4	0.68

a0008 in ARCTIC databases, spoken by male speaker bdl, is converted into 8 different speakers. The original speaker is a tenor, with average pitch of 120 Hz, approximately equivalent to B2 in the music scale, and the MIDI value is 47. Details of the speaker identifications and parameters are shown in Table 8.1. As shown, the first column is speaker identity. The second column is approximate average pitch in Hz. The third column is approximate MIDI value of the average pitch. The fourth column is the music note corresponding to the average pitch. The last column is the relative value of the coefficient in the Laguerre function. The wav files of speech samples are posted on webpage www.columbia.edu/~jcc2161/ under Human Voice. A noticeable fact is that the contrabass and giant voices are deeply in the so-called vocal fry range. Nevertheless, the voice still sounds human, and the vowels and consonants are still clearly identifiable.

8.1.3 Speed change

Using PSOLA, by adding new pitch periods or deleting pitch periods, speed change can be implemented. However, if the speed ratio is too high, the quality deteriorates. One reason for this is that the frames at the border of voiced and unvoiced sections contain vital information, the elimination of those frames would harm the sharpness of the transformed voice.

Using the timbre vector format, the problem can be resolved. First, the set of feature vectors is first segmented into voiced sections and unvoiced sections. The glottal stops and the starting frames of plosives are marked as the starting frames. The speed change is executed by interpolation, similar to the case of a change of pitch contours. Because the critical frames are preserved, the output speech is sharp even at very high speed. In details, if the speed ratio is R, to process a section with a well-defined type starts with frame n_s and ends with frame n_e, the number of interpolated output frames is

$$n_o = (\text{int}) \frac{n_e - n_s - 1}{R}, \tag{8.10}$$

where (int) denotes the integer part of the real number. Define a relative size by

$$S = \frac{n_e - n_s - 1}{n_o}, \tag{8.11}$$

the output feature vectors are calculated as follows. First, the feature vector of the starting frame is copied to the output file,

$$\overline{\mathbf{v}}(0) = \mathbf{v}(n_s), \tag{8.12}$$

where we use a bar to mark the output feature vectors. To compute the interpolated output feature vectors one by one, an index k is introduced,

$$k = 0,\ 1,\ 2,\ \dots n_o - 1. \tag{8.13}$$

A real number τ for the projection of the starting time of the k-th output frame is than computed,

$$\tau = Sk. \tag{8.14}$$

In general, the projected time τ is not an integer. Let the integer part of τ be m,

$$m = (\text{int})\tau, \tag{8.15}$$

a fractional number δ is defined as the

$$\delta = \tau - m; \tag{8.16}$$

the k-th output feature vector is defined as

$$\overline{\mathbf{v}}(k) = (1 - \delta)\mathbf{v}(m) + \delta\mathbf{v}(m + 1). \tag{8.17}$$

After the output feature vectors of all k are written, the last one of the original feature vectors is copied to the output feature vector file,

$$\overline{\mathbf{v}}(n_o - 1) = \mathbf{v}(n_e - 1). \tag{8.18}$$

Examples of the output audio files up to 1000 words per minute are posted on webpage www.columbia.edu/~jcc2161/ under Human Voice.

8.2   Speech Coding

Despite the emergence of video and text modes, since the dawn of telecommunication with the invention of the telephone, voice remains the most natural means of human communication. The analog signals can easily deteriorate during transmission. The full digitized human voice requires a 16 bit PCM at 32 kHz sample rate, or 512 kbit/sec. However, the bandwidth of telecommunication is limited. Therefore, compressing the voice signal into a digital form, transmitting it through the lines or electromagnetic waves, and then recovering the analog voice signal at the receiver side, is the standard method of voice telecommunication [35, 53].

8.2.1   Basic techniques

There are two ways of digitizing and compressing the voice signals. The first technique is to sample the signals at a fixed rate, typically at 8 kHz.

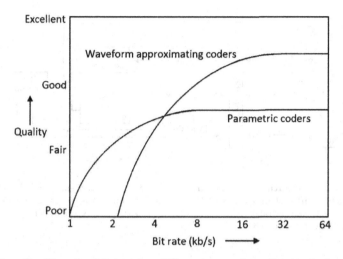

Fig. 8.3. **Quality vs. bitrate for different speech coding techniques.** The waveform approximation coders can provide toll quality output signals at relatively high bitrates. The parametric coders can provide reasonable output signals at low bitrates, typically lower than 4 kb/s. Source: page 6 of Kondoz [53].

The typical raw data at each point is digitized to a 16-bit format. Then the value at each point is compressed to reduce the number of bits to 8. The bitrate, or the number of bits to be transmitted per second is 64 kb/s. That type of compression can produce voice similar to the legacy long-distance telephone calls, which is therefore named *toll quality* [53]. By making further approximations to the waveforms, the bitrate can be further reduced, and the quality further deteriorates, see Fig. 8.3.

Table 8.2: Mean opinion score (MOS) scale
Source: page 17 of Reference [53].

Grade(MOS)	Subjective opinion	Quality
5 Excellent	Imperceptible	Transparent
4 Good	Perceptible, but not annoying	Toll
3 Fair	Slightly annoying	Communication
2 Poor	Annoying	Synthetic
1 Bad	Very annoying	Bad

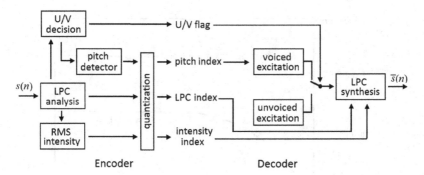

Fig. 8.4. Schematics of a parametric coder. On the encoder side, the input speech signal s(n) first undergoes LPC analysis. The output is quantized into a number of indices. On the decoder side, an LPC synthesizer recovers the speech signal. After Figure 11.1 in Reference [35].

The second technique is to parameterize the speech signal, typically using linear predictive coding (LPC), see Fig. 8.4. As shown, on the encoder side, the speech signal $s(n)$ is first blocked into frames of equal duration and equal shift. The LPC coefficients in each frame is computed, together with a decision of whether the frame is unvoiced or voiced, that is, a U/V decision. If the frame is voiced, a pitch value is detected. The RMS intensity of the frame is also evaluated. Those parameters are then quantized to become digital codes. The codes are than sent to the receiver, that is, the decoder side. According to the U/V flag, if the frame is voiced, a stream of pulses with time intervals of the values of the pitch periods is generated. On the other hand, if the frame is unvoiced, a random noise is generated. The source signal is then processed by the LPC synthesizer to make an output voice signal $\bar{s}(n)$. During the quantization process, the bitrate of the code can be greatly reduced. For LPC coefficients, vector quantization is often used, see the following subsection. In general, by reducing the bitrate, the quality in general further deteriorates, also see Fig. 8.3.

The quality measure in Fig. 8.3 is generated by a subjective test: by sending the original speech signal and the decoded voice signal to a number of listeners, and asking for their opinion. The averaged score, the *mean opinion score* (MOS), is taken as the standard measure of the quality of the data compression scheme. See Table 8.2.

As shown in Fig. 8.3, neither technique could achieve an "Excellent" score. As we have discussed, the full speech signal requires a sample rate of 32 kHz, with 16 bit PCM for each point. The bandwidth is 512 kb/s, which is too high for any telecommunication system. The parametric coders based on LPC never reach a "Good" score even without compression. Nevertheless, at very low bitrates, typically below 4 kb/s, the parametric coders show a definitive advantage over the waveform approximating coders; especially when the technique of *vector quantization* is applied.

8.2.2 Vector quantization

The values of pitch and intensity can be quantized, that is, to be represented by a finite number of levels for transmission. Typically 7-bits, or 128 levels, are used for pitch; and 5 bits, or 32 levels, are used for RMS intensity. The spectral coefficients of speech signals, such as LPC, are always vectors with multiple dimensions. One can quantize each of the dimensions, but the efficiency is low. Vector quantization is often used [33, 49, 53]. Accordingly, the entire vector space is partitioned into a finite number of clusters, each has a center vector and an index. The collection is called a *codebook*, see Fig. 8.5.

During encoding, an input vector is compared with the center vectors in the codebook. The index of the closest matched center vector is transmitted as the index of the input vector. On the decoding side, an exact copy of the codebook is provided. The center vector with the same index is picked up as the output vector, see Fig. 8.5.

The codebook can be built using an automatic procedure, called K-means algorithm, or Lloyd's algorithm [33, 49, 53]. It starts with a collection of vectors $\mathbf{y} \in \mathbf{Y}$, where the vectors must have a distance or distortion measure which satisfies the conditions listed in Section 6.1.

Briefly, the K-means algorithm is as follows [49]:

1. Select K, the number of clusters.

2. Select uniformly at random K initial candidate cluster centers $\mathbf{y}_j^{(0)}$, where $0 \leq j < K$.

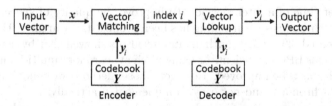

Fig. 8.5. Encoder and decoder with vector quantization. On the encoding side, a feature vector of input speech is compared with a codebook to find a matched vector cluster to produce an index. On the decoder side, a lookup process finds a vector in the codebook according to the index. After Figure 3.9 in Reference [53].

3. Partition the entire space $\mathbf{Y}$ into K clusters $\mathbf{V}_j^{(0)}$, each comprising all vectors nearer to $\mathbf{y}_j^{(0)}$ than to any other $\mathbf{y}_h^{(0)}$, where $h \neq j$.

4. Find the center of gravity of each cluster. For example, if the number of vectors in the j-th cluster $\mathbf{V}_j^{(0)}$ is $n_j^{(0)}$, the center of gravity is

$$\mathbf{y}_j^{(1)} = \frac{1}{n_j^{(0)}} \sum_{\mathbf{y}_i \in \mathbf{V}_j^{(0)}} \mathbf{y}_i. \tag{8.19}$$

5. Repartition $\mathbf{Y}$ into into K clusters $\mathbf{V}_j^{(1)}$, each comprising all vectors nearer to $\mathbf{y}_j^{(1)}$ than to any other $\mathbf{y}_h^{(1)}$, where $h \neq j$.

6. Find the center of gravity $\mathbf{y}_j^{(2)}$ of each cluster $\mathbf{V}_j^{(1)}$.

7. Continue until the cluster centers converge; that is, until the sum of the differences of $\| \mathbf{y}_j^{(n+1)} - \mathbf{y}_j^{(n)} \|$ is smaller than a given value.

For more details of vector quantization, see Gersho and Gray [33]. Using vector quantization with LPC vocoder, the bitrate of a codec system can be further reduced to around 1 kb/s. As we have discussed above, the quality of an LPC vocoder system is worse than the toll quality even without quantization. With vector quantization, understandably, the quality further deteriorates [35, 53].

8.2.3   Coding based on timbre vectors

The mediocre performance of the codec systems based on LPC stems from its non-pitch-synchronous nature. It is well known that the voiced speech signal is pseudo-periodic, and the LPC coefficients become inaccurate at the onset time of a pitch period. To improve the quality of speech coding, pitch-synchronous speech coding has been proposed, researched [94, 101, 89] and patented [30]. Those authors and inventors showed that by using pitch-synchronous LPC coefficients or using pitch-synchronous multi-band coding, the quality can be improved. However, those methods were not thoroughly pitch-synchronous, and not easy to implement practically.

Because the quality of a timbre vector based vocoder can be better than toll quality, with vector quantization, the quality of a timbre vector based low-bitrate codec system can be better than LPC-based codec systems. Figures 8.6 through 8.8 show an example [13].

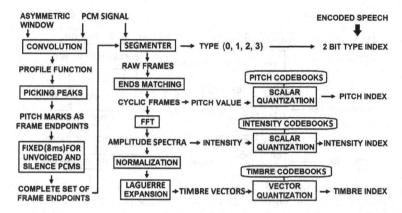

Fig. 8.6. Encoder based on timbre vectors. The PCM is first segmented into pitch-synchronous frames, to generate a timbre vector, which is then vector-quantized using a timbre codebook to generate a timbre index. After Reference [13].

Besides thoroughly pitch-synchronous, another advantage of the timbre vectors is a well-behaved distance measure, or distortion measure,

$$d(\mathbf{x}, \mathbf{y}) \equiv \parallel \mathbf{x} - \mathbf{y} \parallel = \sqrt{\sum_{n=0}^{n<N} (x_n - y_n)^2}, \qquad (8.20)$$

which is applied to the normalized Laguerre spectral coefficients.

Figure 8.6 shows an encoder based on timbre vectors. The input signal, typically in PCM (pulse-code modulation) format, is first convoluted with an asymmetric window, to generate a profile function. The peaks in the profile function, with values greater than a threshold, are assigned as pitch marks of the speech signal, which are the frame endpoints in the voice section of the input speech signal. If no peaks in the profile function are found in a 32-msec buffer, an unvoiced frame is assigned with a fixed frame size, typically 8 msec. Therefore, the frame endpoints are generated on the run. Using those frame endpoints, the PCM signal is segmented into raw frames. In general, the PCM values of the two ends of a raw frame do not match. By performing Fourier analysis on those raw frames, artifacts would be generated. An ends-matching procedure is applied on each raw frame to convert it into a cyclic frame which can be legitimately treated as a sample of a continuous periodic function. Then, a fast Fourier transform (FFT) unit is applied to each cyclic frame to generate an amplitude spectrum. The intensity of the spectrum, the RMS value, is calculated as the intensity value. The amplitude spectrum is then normalized, and is then expanded

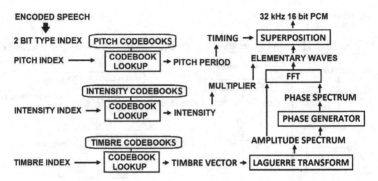

Fig. 8.7. Decoder based on timbre vectors. based on the incoming indices, a look-up procedure using the codebooks generates timbre vectors, pitch and intensity. The speech signal is then recovered. After Reference [13].

using Laguerre functions, to generate a set of expansion coefficients, which constitute a timbre vector.

To reduce delay time, the above process is executed on a frame-by-frame basis. A buffer of 32 msec is retrieved from the speech signal stream. A profile function is computed using an asymmetric window. Once a peak of the profile function is found, a voiced frame is identified. The buffer is advanced by the length of that frame. If no peak is found inside the 32 msec segment, a 8-msec segmentation point is set up, the frame is identified as unvoiced, and the buffer is advanced by 8 msec.

During the above process, the type of the frame (pitch period for voiced frames) is determined. If the amplitude is smaller than a silence threshold, the frame is silence, type 0. If the intensity is higher than the silence threshold but there is no pitch marks, the frame is unvoiced, type 1. For frames bounded by pitch marks, if the amplitude spectrum is concentrated in the low-frequency range (0 to 5 kHz), then the period is voiced, type 3. If the amplitude spectrum in the higher-frequency range (5 to 16 kHz) is substantial, for example, has 30% or more power, then the period is transitional which is a voiced fricative or a transition frame between voiced and unvoiced, type 2. The type information is encoded in a 2-bit type index. For voiced periods, the pitch value is conveniently expressed in MIDI unit. Using a pitch codebook, the pitch is scalar-quantized. The intensity is conveniently expressed in decibel (dB) unit. Using an intensity codebook, through scalar quantization, the intensity index of the frame is generated. Furthermore, using a timbre codebook generated through the K-means algorithm, using vector quantization, the timbre index of the frame is generated. Notice that for each type of frame, there is a different codebook. The four types of indices are then transmitted to the receiver end.

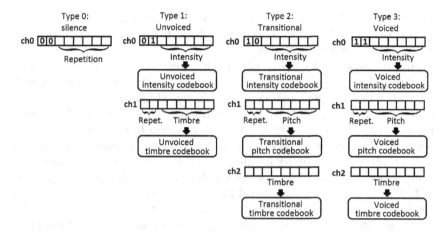

Fig. 8.8. Data structure of a codec system based on timbre vectors. A simple scheme using a finite number of bytes. For different types of frames, the number of bytes and the bit allocation are different. After Reference [13].

Figure 8.7 shows the decoding process. From the signals transmitted to the decoder, the 2-bit type index is first fetched. If the frame is silence, a silence PCM, 8 msec of zeros, is sent to the output. If the frame is voiced, type 3, or transitional, type 2, the pitch index, the intensity index, and the timbre index are fetched. Using the pitch codebook for voiced frames or the pitch codebook for transitional frames through a look-up procedure, the pitch period is identified. Using the intensity codebook for voiced frames or the intensity codebook for transitional frames, through another look-up procedure, the intensity of the frame is determined. The intensity value and the timbre vector are then sent to the waveform recovery unit to generate the elementary wave for that frame. The procedure is as follows. First, using Laguerre transform, the timbre vector is converted back to amplitude spectrum. Then, using a phase generator based on Kramers-Kronig relations, the phase spectrum is generated from the amplitude spectrum. Using fast Fourier transform (FFT), an elementary waveform, that is, a timbron, is generated. In order to generate PCM output, those elementary waves are lineally superposed according to the time delay defined by the duration of each pitch period. For unvoiced frames, type 1, the procedure is identical, except the frame duration is a fixed value, typically 8 msec; and the phase spectrum is random over the entire frequency scale.

Figure 8.8 shows a possible simplified scheme of bit allocation. The design is a proof-of-the-concept coding scheme, not optimized for quality and minimizing the bandwidth. In that design, only integer number of bytes is used. Therefore, it can be viewed by displaying the octal values of each byte. In that design, the number of the frame repetition is encoded, represented by a repetition index, see below.

If the first byte of a group of bytes has highest bits of 01, the frame is unvoiced. The frame duration is also 8 msec. Pitch index is not required. The rest 6 bits are the intensity index. By consulting an unvoiced intensity codebook, the intensity of that unvoiced frame is determined. Each unvoiced frame is represented by two bytes. The first two bits of the second byte represent the number of repetition. If two consecutive frames have similar timbre vectors, the repetition index is 1. If three consecutive frames have similar timbre vectors, the repetition index is 2. The maximum repetition is 3. This upper bound is designed for two purposes. First, the intensity of the repeated frames has to be interpolated from the end-point frames. To ensure quality, a limit of four frames is needed. Second, the encoding of four repeated unvoiced frames takes 32 msec. Because the tolerable encoding delay is 70 to 80 msec, 32 msec is acceptable. On the other hand, too many frames would cause too much encoding delay.

If the first two bits of the leading byte are 10 or 11, the frame is voiced or transitional, and two following bytes should be fetched from the transmission stream, ch1 and ch2. As in the case of unvoiced frames, the remaining 6 bits of the leading byte represent the intensity index. By consulting an intensity codebook, the intensity is determined. The second byte carries a repetition index and a pitch index. The repetition index is limited to 3, and both intensity and pitch have to be linearly interpolated from the two ending-point frames. By consulting a pitch codebook, the pitch value is determined. The third byte is timbre index. By consulting a timbre codebook, the timbre vector is determined. Because each type of frame is distinct, a codebook size of 256 for each type seems adequate. During encoding, the determination of type 2 (transitional) and type 3 (voiced) is based on the spectral distribution, as presented above. If the speech power in a frame with a well-defined pitch period is concentrated in the low-frequency range (0 to 5 kHz), the frame is voiced. If the power in the high frequency range (5 kHz and up) is substantial, then it is a transitional frame. During encoding, different types of frames are treated differently. For voiced frames, below 5 kHz, the phase is generated by the Kramers-Knonig relations; and above 5 kHz, the phase is random. For transitional frames, below 2.5 kHz, the phase is generated by the Kramers-Knonig relations; and above 2.5 kHz, the phase is random. For unvoiced frames, the phase is random on the entire frequency scale.

To improve speed, fast Fourier transform (FFT) is applied. However, FFT is much more efficient if the period is an integer power of 2, such as 64, 128, 256, etc. For voiced frames, the pitch period is a variable. In

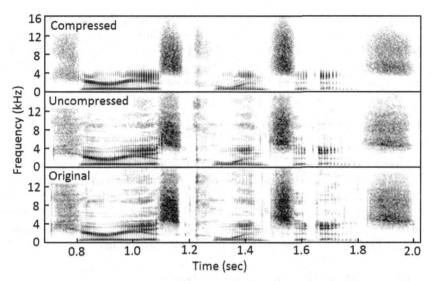

Fig. 8.9. An example of vocoded speech. Using a very low bitrate coding, near CD quality speech signals can be achieved. The original sentence is a0052 in ARCTIC databases, spoken by a male speaker bdl. Note that the frequency scales of voiced and unvoiced frames are set differently, to reduce computational load.

order to utilize FFT, the PCM values in each pitch period are first linearly interpolated into 2^n points, typically 256 points. After FFT, the amplitude spectrum is reversely interpolated to the true values of the pitch period. To reduce computation load, the dimension of the timbre vector is set to 12. For voiced, transitional and unvoiced frames, different scale constants κ are chosen. For voiced frames, the frequency range is limited to 0 – 5 kHz. For unvoiced frames, the frequency range is 0 – 16 kHz, and the frequency resolution is about 4 times lower, for which a low frequency resolution is acceptable. The frequency range and resolution of transitional frames are set to be intermediate.

An example of the vocoded speech is shown in Fig. 8.9, encoded from sentence a0052, spoken by a U.S. English speaker bdl, in ARCTIC databases [52]. The duration of the speech is 2.5 seconds. The encoded speech has 543 bytes, or 4344 bits. Therefore, the bit rate is 4.344/2.5=1.737 kb/s, in the very-low-bit-rate range. Nevertheless, nearly CD-quality voice is regenerated. The advantages of the current method are predicable from its principle. First, the maximum bandwidth of the speech signal can be 16 kHz or greater using a PCM speech signal of 32 kHz sampling rate or higher and 16 bit resolution. The legacy speech coding is based on 4 kHz bandwidth (8 kHz PCM sampling rate, 8 bit), the fricatives such as [f] and [s] are not distinguishable. Using the algorithm presented here, the fricatives [f] and [s] are clearly distinguishable. Furthermore, while the legacy low-bit-rate speech coding is based on an all-pole model of speech signal that fails to represent the nasal sounds, the technology presented here reproduces the entire spectrum, and the nasal sounds are reproduced faithfully.

8.3 Speech Recognition

A speech recognizer is a device that automatically transcribes speech into text [49]. It has two components. The *frontend*, or acoustic processor, converts the spoken speech into a series of phonetic symbols. The *backend* converts the phonetic symbols into text. The purpose of this Section is to introduce a pitch-synchronous signal processing method based on timbre vectors for improving the accuracy of acoustic processing.

8.3.1 Frontend of a speech recognizer

A schematic diagram of the frontend of a speech recognizer is shown in Fig. 8.10 [49]. The speech signal first goes through an signal processor to produce a series of vectors $\mathbf{x}_0$, $\mathbf{x}_1$, $\mathbf{x}_2$, ... $\mathbf{x}_i$, ... which represent the acoustic characteristics of the speech signal. Each vector is compared with the prototype vectors in a prototype storage. In the prototype storage, each prototype vector is associated with a phonetic symbol α_j in a phonetic alphabet A. With the comparator, the phonetic symbol α_i associated with the closest vector $\mathbf{y}_{\alpha_i}$ in the prototype storage is produced.

The phonetic alphabet typically follows the standard IPA alphabet. For US English, following the Sphinx phone system of CMU, there are 39 phones, which can be classified into several types as follows:

- **Monophthongs**. The timbre of the phone is basically invariant over the entire duration, including AO, AA, IY, UW, EH, IH, UH, AH, AX, and AE. In certain dialects, ER also belongs to this category.

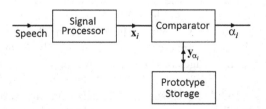

Fig. 8.10. A schematic diagram of the frontend of a speech recognizer.

- **Diphthongs**. The timbre of the phone shifts during the duration of the phone, including EY, AY, OW, AW, and OY.

- **Fricatives** It is an audible continuous friction without changing of timbre, including F, V, TH, DH, S, Z, SH, and DH.

- **Stops** It contains a nasal or a silence moment followed by a plosive of air, including P, B, T, D, K, and G.

- **Affricates** It contains a nasal or a silence moment, a plosive followed by a fricative, including CH and JH.

- **Nasals** Voiced consonants with mouth closed, including M, N, and NG.

- **Liquids** Pointing to voiced consonants including L and R.

- **Semivowels** Shortened and weakened vowels before or after a main vowel, including Y and W.

As shown, some phonemes have a constant timbre over the entire duration, but three types of phonemes have internal structure: A diphthong is a combination of two vowels. A voiced stop contains a nasal and a plosive. An unvoiced stop contains a silence moment and a plosive. A voiced affricate contains a nasal, a plosive, and a fricative. An unvoiced affricate contains a silence moment, a plosive, and a fricative.

8.3.2   Acoustic processor with a fixed window size

Since the middle of the 20th century, the method of fixed windows has been used extensively for speech signal processing [6, 25, 32, 34, 69, 70]. Typically, the speech signal is blocked into frames with a duration of 25

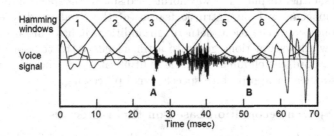

Fig. 8.11. Pitch-asynchronous segmentation. The speech signal is blocked into frames with 10 msec shift and 25 msec duration, and multiplied with a window function, typically a Hamming window. The frames are asynchronous to the underlying pitch period structure, always cross the pitch-period boundaries, and often even cross the boundaries between voiced signals and unvoiced signals, see points A and B.

msec and a shift of 10 msec, then multiplied with a window function. A popular window function is the Hamming window [70],

$$h(n) = 0.54 - 0.46\cos(2\pi n/(N-1)), \quad 0 \le n \le N - 1, \qquad (8.21)$$

otherwise $h(n) = 0$. The windows duration is 2.5 times greater than the window shift, making the adjacent windows overlap significantly, see Fig 8.11. As shown in Fig. 8.11, because the windows are asynchronous to the underlying pitch periods, in the voiced sections of speech signal, every window contains at least two pitch boundaries. Pitch information and timbre information are not separated. Many windows contain boundaries of a voiced section and an unvoiced section, as shown in frames 3, 4, 5, and 6 at point A and B in Fig. 8.11. Therefore, the information in most framed speech signals inevitably contain mixed phonemes of different types.

As we have presented previously, many phonemes are not a single entity, but a concatenation of two or three distinct pieces. Therefore, in general, to take the phonemes as acoustic symbols is not feasible. A general practice is to divide a phoneme into three parts, the starting part, the middle part, and an ending part. For example,

AX → AX_1, AX_2, AX_3.
AW → AW_1, AW_2, AW_3.
F → F__1, F__2, F__3.
K → K__1, K__2, K__3.
CH → CH_1, CH_2, CH_3.

Notice that using the traditional fixed-frame-size segmentation method, a frame often crosses the boundary of two phonemes; thus the identification of the phonetic symbols with an acoustic vector is often mixed. This situation becomes very clear from inspecting records of automatically aligned voice files, either displayed as waveforms or displayed as spectrograms. This type of inaccuracy is an intrinsic feature of the fixed window or pitch-asynchronous segmentation method. In addition, it has long been known that all windows functions generate artificial effects to the speech signal [37]. Therefore, the elimination of the multiplication of processing windows provides a cleaner approach to speech signal processing.

8.3.3   Speech recognition based on timbre vectors

As an application of the mathematical methods presented in Chapter 5, 6, and 7, a novel method of acoustic processing in speech recognition systems can be established [14]. The frames are pitch-synchronous, therefore, none of the frames can cross a phoneme boundary. In other words, each frame should have a definitive phoneme identity, including transitional phonemes.

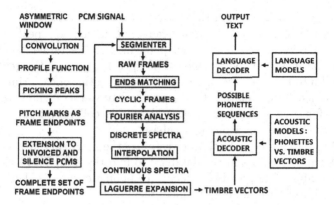

Fig. 8.12. Speech recognition system based on timbre vectors. The speech signal goes first through a pitch-marks picking program to generate segmentation points. The speech signal is segmented into pitch-synchronous frames. An ends-meeting program is executed to make the values at the two ends of every frame equal. Using Fourier analysis, the speech signal in each frame is converted into a pitch-synchronous amplitude spectrum. Then, using Laguerre functions, it is converted into a unit vector characteristic to the instantaneous timbre, the timbre vector.

The elimination of a process window removes the artifacts due to the window functions. A schematic diagram is shown in Fig. 8.12.

As shown in Fig. 8.12, the speech signal in PCM format is first convoluted with an asymmetric window, to generate a profile function. The peaks in the profile function, with values greater than a threshold, are assigned as pitch marks of the speech signal, which are the frame endpoints in the voice section of the input speech signal. The pitch marks only exist for the voiced sections of the speech signal. Those frame endpoints are then extended into unvoiced and silence sections of the PCM signal, generating a complete set of frame endpoints. Through a segmenter, using those frame endpoints, the PCM signal is segmented into raw frames. In general, the PCM values of the two ends of a raw frame do not match, and performing Fourier analysis on those raw frames would create artifacts. An ends-matching procedure is applied on each raw frame to convert it into a cyclic frame which can then be legitimately treated as a sample of a continuous periodic function. Then, Fourier analysis is applied to each cyclic frame to generate an amplitude spectrum. The amplitude spectrum is expanded using Laguerre functions, to generate a set of expansion coefficients, then normalized to become a timbre vector.

As we have discussed in Chapter 5, the polarity of the asymmetric window is determined by the speech acquisition system, and the width of the asymmetric window depends on the speaker. This is periodically updated using the running average of the pitch values of the speaker.

Using recorded speech by a speaker or a number of speakers reading a prepared text which contains all phonemes of the target language, an acoustic database can be formed. The speech signal of the read text is converted into timbre vectors. The acoustical identity of each timbre vector is determined by correlation to the text. The average timbre vector and variance for each individual acoustical unit is collected from the paired record, which forms an acoustic database.

Similarly to Fig. 8.1, the timbre vectors of the input speech are compared with the timbre vectors in the database to find the most likely sequence of phonetic units. The possible sequence of phonetic units is then sent to a language decoder to find out the most likely text.

8.3.4   Concept of phonette

In the previous subsection, a vague term "phonetic unit" is used instead of "phoneme" in the conventional presentation. This is because of the nature of the timbre vectors.

First, let us make a general remark on the terminologies. In the linguistic community, *phoneme* and *phone* are two distinct concepts. Phoneme is the minimal unit in the sound system of a language, a subject of *phonology*, enclosed by a pair of slashes, such as /k/. Phone is the physical realization of phoneme, a subject of phonetics or *acoustic phonetics*, enclosed in a square bracket, such as [k]. Allophones are different phones by which an identical phoneme can be realized [68, 20, 22, 59]. Nevertheless, the size of a phone is identical to the size of the phoneme it represents.

As presented in Section 8.3.1, many phonemes have internal structures. For example, the diphthong AW starts with a vowel AA and ends with a vowel WU. However, each timbre vector only lasts a single pitch period. Therefore, at the beginning of the diphthong AW, the timbre vector resembles that of AA; and at the end of the diphthong AW, the timbre vector resembles that of UW. A stop starts with a moment of silence, then a plosive occurs, followed by a noise section characteristic of the stop consonant. The timbre vector of a plosive resembles that of a vowel, which has a phase spectrum determined by its amplitude spectrum.

Therefore, in the timbre-vector format, where each pitch period is a unit, a phonetic symbol system of a lower layer than the traditional phone layer should be established. Although in many cases, such as the monophthongs and fricatives, a phoneme corresponds to a timbre vector, in many cases, a phoneme must be decomposed into several smaller phonetic units, each is called a phonette. The word "phonette" originally means a cellular phone as a smaller version of a telephone device. Here we borrow that word to indicate a phonetic unit as a smaller version of a phone.

To implement speech recognition systems based on timbre vectors, the phonetic dictionary for speech recognition should be refined into a list of phonettes. Furthermore, in many languages, the long vowel and the short vowel, with the same timbre, are different phonemes. The difference between aspirated plosives and the conventional unvoiced plosives is the length of weak noise section between the plosive and the following phoneme. Therefore, the timing is a factor in the phonetic transcription process.

On the other hand, to characterize the diphthong more precisely, a dynamic phonette can be introduced. It is represented by a differential timbre vector as the difference of the two adjacent timbre vectors. In a conventional speech recognition system, the differentials of the feature vectors are included in the acoustic process. Therefore, the inclusion of dynamic phonette can be natural.

8.4 Speech Synthesis

Speech synthesis, or text-to-speech (TTS) synthesis, is to convert text into speech using computer software, has found wide applications in telecommunications, entertainment, and education [81]. The system has two components. The TTS *frontend* processes the text, producing a phonetic sequence with prosodic information, including pitch contour, intensity profile, and duration profile. The TTS *backend* produces speech sound according to the information provided by the frontend.

In this Section, we present methods using the timbre vector parameterization to develop the TTS backend, improving the accuracy, intelligibility, and naturalness of the synthesized speech.

8.4.1 Formant synthesizer and concatenative synthesizer

For several decades, the two most important TTS synthesizer backend technologies have been the formant synthesizer and the concatenative synthesizer.

The formant synthesizer, a kind of rule-based TTS synthesizer, is based on the source-filter theory of speech production. For voiced sounds, the source is the periodic glottal flow function, and the filter, representing the vocal tract, is formulated as a series of formants with central frequency, strength, and width. For unvoiced consonants, the source is white noise. Properly designed formant TTS synthesizers can produce accurate and intelligible speech using very limited computing resources. Furthermore, the output is highly versatile: the speaker identity (man, woman, older or younger), the speed, and the speaking style can be set up at will. However, the sound is characteristically robotic [81].

The concatenative TTS synthesizer represents another extreme. A huge inventory of speech is recorded. During synthesis, the phonetic and prosodic input is chopped into small pieces, a searching mechanism looks for matched fragments of speech from the huge pool of recorded speech, then concatenates those small pieces together. This gives it an alternative name of unit-selection TTS synthesizer. However, the joints between the units often cause a problem of discontinuity. Only very limited signal processing can be used, such as the pitch-synchronous overlap-and-add technique (PSOLA) [9, 65, 66]. Nevertheless, if the size of the body of recorded speech is large enough, matched units with smooth joints can be found. Because it is almost like playing back a recorded speech, naturalness is guaranteed. However, in contrast to the formant TTS synthesizer, the sound is exactly the voice of the recording speaker, and it is almost impossible to change speaker identity or speaking style [81]. Furthermore, a huge storage space is required.

A perennial research direction in TTS technology is to find a synthesizer that is as versatile and accurate as the formant synthesizer, and as natural sounding as the unit-selection synthesizer [96, 104].

8.4.2   Speech synthesizer based on timbre vectors

As we have shown, both the formant synthesizer and the concatenative synthesizer have shortcomings, and the shortcomings originate from the same source: neither method is based on an accurate understanding of the physics of human voice production. The formant synthesizer is based on an oversimplified theory of human voice production: the source-filter model. The parameters are not obtained from real speakers, but are theorized. The concatenative synthesizer goes to another extreme: no theory of human voice production is applied. Just copy, copy, and copy.

In Part I of this book, the timbron theory of human voice is presented. In Chapters 5, 6, and 7, mathematical representations of human voice based on the timbron theory are provided. If the timbron theory correctly represents how human voice is produced, a good TTS synthesizer could be constructed by cloning the natural process of human voice production.

Figure 8.13 shows a schematic diagram of a TTS system based on timbre vectors. Basically, the recorded speech is disassembled into its elementary components. Those elementary components, timbrons, are then properly parameterized. During synthesizing, those elements of voice, similar to atoms, are reassembled according to the instructions from the frontend, similarly to following a chemical formula.

The speech synthesis system has two major parts: a database building

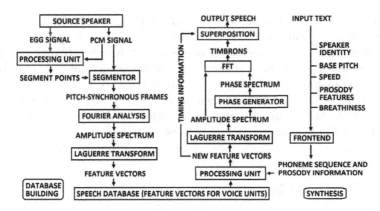

Fig. 8.13. Speech synthesis based on timbre vectors. The left-hand side shows database construction. Recorded speech is segmented into pitch-synchronous frames, then a feature vector for each frame is generated. The unit identity and the corresponding series of feature vectors are grouped to form the TTS database. The right-hand side shows the process of TTS synthesis. The input text first goes through the frontend, to generate a set of phonemes with prosody features. The timbre vector parameterization enables full control of timbre and prosody. A variety of effects can be created [11].

part (the left-hand side of Fig. 8.13), and a synthesizing part (right-hand side of Fig. 8.13). In the database building unit, a source speaker reads a prepared text. The voice is recorded by a microphone to become the PCM signal. The optional glottal closure signal is recorded by an electroglottograph (EGG) to become EGG signal. The EGG signal and the PCM signal are used by the processing unit to generate a set of segment points. Then the PCM signal is segmented by the segmenter into frames using the segment points. Each frame is processed by a unit of Fourier analysis to generate an amplitude spectrum. The amplitude spectrum of each frame is then expanded using Laguerre functions, and normalized to a unit vector, which represents the instantaneous timbre of that frame, that is, a timbre vector. Together with the voicedness index, duration and intensity parameter, a feature vector is formed. The collection of feature vector and its corresponding speech unit, for example, phonemes, diphones, demisyllables, syllables, words and even phrases, are then stored in the speech database.

In the synthesis unit the input text, together with synthesis parameters, is fed into the frontend. The frontend generates detailed instructions about the phonettes, intensity and pitch values, which are then transferred to a processing unit. The processing unit selects timbre vectors from the database, converting them to a new series of timbre vectors according to the instructions from the process unit, using timbre fusing if necessary (see

Section 8.4.3). Each timbre vector is converted into an amplitude spectrum using Laguerre transformation. The phase spectrum is generated from the amplitude spectrum by a phase generator using Kramers-Kronig relations. The amplitude spectrum and the phase spectrum are sent to a FFT (Fast Fourier Transform) unit, to generate an elementary acoustic wave. The elementary acoustic waves are then superposed by the superposition unit according to the timing information provided by the new feature vectors, generating the final result, the output speech signal.

8.4.3   Timbre fusing

One of the difficulties in the concatenation TTS synthesizer is the discontinuity at the border of two segments of voice. Using timbre vector parameterization, there is a natural solution. By making a weighted sum of the input timbre vectors across the juncture, the entire series of frames can be fused into a smooth unit with a continuous transition [11].

Figure 8.14 shows schematically the principle of timbre fusing. The first three input timbre vectors belong to phone S_1, and the next three input timbre vectors belong to phone S_2. The output timbre vector A' is the same as input timbre vector A. The output timbre vectors B', C', D' and E' are the weighted sum of the input timbre vectors. In general, the weighted sum can go much further than that in Fig. 8.14.

Suppose there are $2N$ timbre vectors, v_0, v_1, v_2, ... v_n, ... , with first N frames belong to phone S_1 and the last N frames belong to phone S_2. For $0 \leq k < N$, define the output timbre vectors to be

$$\mathbf{v}'_k = \frac{(2k)!}{2^{2k}} \sum_{n=0}^{n=2k} \frac{\mathbf{v}_n}{n!(2k-n)!}, \qquad (8.22)$$

Fig. 8.14. Principle of timbre fusing. The first three timbre vectors belong to phone S_1, and the next three timbre vectors belong to phone S_2. By making a weighed sum of the input timbre vectors to generate a set of output timbre vectors, a smooth transition can be generated [11].

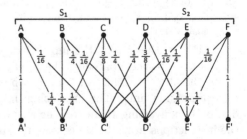

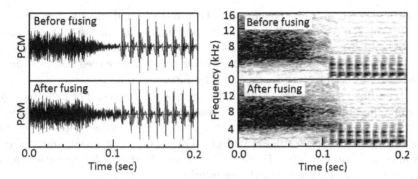

Fig. 8.15. Effect of timbre fusing. Butt-joining an [s] with an [ɑ] creates a discontinuity. After fusing with $N = 5$, a smooth transition is generated [11].

and for $N \le k < 2N$,

$$\mathbf{v}'_k = \frac{(4N - 2k - 2)!}{2^{4N - 2k - 2}} \sum_{n=2k-2N+1}^{n=2N-1} \frac{\mathbf{v}_n}{(2N - n - 1)!(2N - 2k + n - 1)!}.$$

(8.23)

The case on Fig. 8.14 represents $N = 3$.

The averaging can be extended to the entire feature vector, including voicedness index, duration of the pith period, and RMS intensity. In this case, it is actually a *feature fusing*.

Figures 8.15 and 8.16 show two examples of the effect of timbre fusing. In Fig. 8.15, a consonant [s] followed by a vowel [ɑ], and the two are concatenated together. Before fusing, on both waveforms and spectrogram, a sharp discontinuity is apparent. After fusing with $N = 5$ across the boundary, a

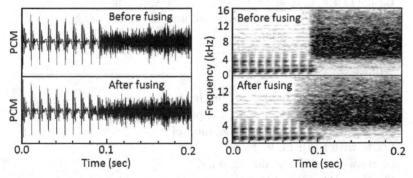

Fig. 8.16. Effect of timbre fusing. Butt-joining an [ɑ] with an [s] creates a discontinuity. After fusing with $N = 5$, a smooth transition is generated [11].

continuous transition is generated. The waveform and the spectrogram after fusing look very similar to the natural transition in authentic human speech. In Fig. 8.16, the reverse case is demonstrated: a vowel [ɑ] followed by a consonant [s]. Again, a sharp discontinuity is observed before fusing, and a smooth transition is generated by fusing.

The fusing procedure may create unexpected effects. For example, by fusing [fin] and [gə] together, the consonant [n] is simulated to [ŋ]. By fusing [sin] and [bʌl] together, the [n] is simulated to [m]. Furthermore, not every fusing operation creates desired effects. For example, by fusing [ɑ] with [n], it always sound like [ɑm]. However, it is a simple and useful operation, and more experiments are worthwhile exploring.

8.4.4   Irreducible units

In traditional concatenative TTS synthesizers, huge number of units must be stored, because the difficulties in voice transformation and the discontinuities at the joints. The timbre vector parameterization enables large-scale voice transformation while preserving naturalness, reducing the required number of units dramatically. Furthermore, by applying the timbre fusing algorithm, the number of units required to generate smooth and natural speech can be further reduced.

Take an example of Mandarin Chinese. Because tones, or pitch contours, can be easily implemented using voice transformation, see Section 8.1.1, and the intersyllable joints can be smoothed out using timbre fusing, only the 400 raw syllables without tones are needed for a full Mandarin TTS synthesizer. Furthermore, using timbre fusing, with allophones of initials and all finals [102], the irreducible units are as follows:

Finals:

Basic (11): a, o, e, ai, ei, ao, ou, ang, en, eng, ong
Starts with u (9): u, ua, uo, uai, ui, uan, uang, un, ueng
Starts with i (10): i, ia, ie, iao, iu, ian, iang, in, ing, iong
Starts with ü (4): ü, üe, üan, ün

Allophones of initials:

Starts with b (5): ba, bo, be, bu, bi
Starts with p (5): pa, po, pe, pu, pi
Starts with m (5): ma, mo, me, mu, mi
Starts with f (4): fa, fo, fe, fu
Starts with d (5): da, do, de, du, di
Starts with t (5): ta, to, te, tu, ti
Starts with n (6): na, no, ne, nu, ni, nü

Starts with l (6): la, lo, le, lu, li, lü
Starts with g (4): ga, go, ge, gu
Starts with k (4): ka, ko, ke, ku
Starts with h (4): ha, ho, he, hu
Starts with z (5): za, zo, ze, zu, zi
Starts with c (5): ca, co, ce, cu, ci
Starts with s (5): sa, so, se, su, si
Starts with sh (5): sha, sho, she, shu, shi
Starts with zh (5): zha, zho, zhe, zhu, zhi
Starts with ch (5): cha, cho, che, chu, chi
Starts with r (5): ra, ro, re, ru, ri
Starts with j (2): ji, jü
Starts with q (2): qi, qü
Starts with x (2): xi, xü

The total number of units is reduced from 400 to 128.

8.4.5. Phonette-based universal TTS synthesizer

According to the International Phonetic Association [68], a large number
of vowels in various languages can be marked on a two-dimensional vowel
chart. The horizontal axis is from Front to Back, and the vertical axis is
from Open to Close, see Fig. 8.17. Each point on the chart can be associated
with a phonette, that is, a timbre vector. By asking a good speaker to record
a large number of points densely located on the vowel chart, those points
can be arranged as a two-dimensional array, that is, a chart of phonettes
labeled with two indices. Thus each phonette, as well as each timbre vector, is labeled by two indices. To have a more accurate representation, a
third index or even a fourth index may be needed to indicate, for example,
rounded vowels, and nasalized vowels. To synthesize a diphthong, one can
draw a line in the vowel chart, and output a series of timbre vectors to
generate that diphthong, see Fig. 8.17. Triphthongs can also be generated

Fig. 8.17. The IPA vowel chart.
A large number of vowels in various
languages can be marked on a two-
dimensional vowel chart. Each represents a phonette, and therefore a
timbre vector. A sequence of timbre vectors constitutes a diphthong.
A phonette-based universal TTS synthesizer can be constructed. Adapted
from IPA Handbook [68].

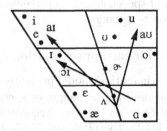

using the trajectory method. Such a database may be used to construct a universal TTS synthesizer for most languages in the world.

Note that under the phonette scheme, monophthongs, diphthongs, triphthongs, glides, and semivowels are treated on the same footing.

8.5   Syllable-Centered Pitch Parameterization

Besides timbre, which determines the identity of vowels and consonants, an equally important aspect of human speech is *prosody*, represented by acoustic features that include pitch contour, relative intensity, and duration. More than one half of world's languages are tone languages [103], which use pitch contours of the syllables to distinguish words or their inflections, making pitch variation an essential element of speech signals.

Generating acoustic prosody features is a significant subfield in TTS technology. For a review up to 2007, see Chapter D.33 of Springer Handbook for Speech Processing, *Prosodic Processing* [6]. Broadly speaking, there are two commonly used approaches. In the early stage of TTS technology, rule-based prosody generation works side by side with rule-based timbre generation, such as in the Klatt TTS system. In the later unit-selection TTS systems, to preserve naturalness of the speech, prosody modification is seldom applied. For a phrase with the same words, a number of versions with different prosody features are recorded. During speech synthesis, the most closely matched prosodic version of the recorded phrase is selected. Apparently, to cover arbitrary prosodic variations, tremendous numbers of recorded speech samples are needed.

The parameterization methods presented in Chapters 5, 6, and 7 enable dramatic prosodic modifications of a piece of recorded speech without losing naturalness. Therefore, to synthesize speech with arbitrary prosody, a much smaller inventory of recorded speech samples is required. To take advantage of the versatility of the parameterization using timbre vectors, a more efficient mathematical model of prosody is needed. In this Section, a mathematical model of prosody suitable for the timbre vector parameterization of speech signals is presented. It can be used for both tone languages and non-tone languages.

8.5.1   Linear approximation

The mathematical model presented here is inspired by the parameterization method used in tone language speech recognition, characterized by the concept of *tonemes* [16, 18]. Accordingly, the tone information in each syllable is carried by the pitch value, the time derivative of pitch, and the second

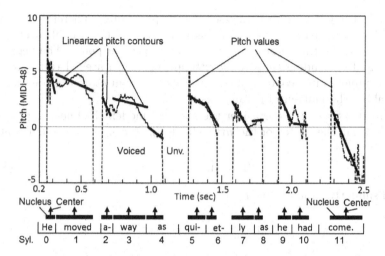

Fig. 8.18.   Linear approximation of pitch contours. The pitch contour of the voiced sections of each syllable is approximated by a linear function with two coefficients, and those two coefficients are obtained by least-squares fitting [10].

time derivative of pitch at the center of the main vowel in a syllable. The pitch curve from the center of one syllable to the center of the next syllable is determined by the condition of smooth continuation with least effort. Such a concept is based on the process of tone generation: The pitch values and their derivatives at the centers of syllables follow the order of the central nervous system, as well as the features which can be easily sensed, because the speech power is strong at the centers of syllables. Those pitch values and their derivatives determine the tone of the main vowel in a syllable. The vowels with different tones are different phonemes, called tonemes.

Following that concept, the parameters required to represent the entire pitch contour of an utterance are the pitch values and the pitch derivatives at the centers of the syllables. For non-tone languages, the pitch value and the first derivative of pitch at the center of each syllable are often sufficient to represent the entire pitch contour.

The concept of syllable-centered parameterization of pitch contours is shown schematically on Figs. 8.18 and 8.19, taking an example of sentence a0045 in ARCTIC databases, spoken by a US English male. For each syllable, the nucleus is identified. (Automatic identification of syllable nucleus is often difficult. Practically, the voiced portion of a syllable, which is relatively easy to identify automatically, is used as the syllable nucleus.) The midpoint of the nucleus is defined as the center of the syllable. For each syllable, a linear approximation of the pitch contour on its nucleus is calculated using a weighted least-squares fit procedure. Therefore, for each syllable, three parameters are needed: the time of the syllable center, the pitch, and the time derivative of pitch at the center point. The beginning time and the ending time of the utterance are also needed to reconstruct the complete pitch contour of the utterance. For an utterance containing n syllables, the number of parameters is $3n + 2$.

In this Section, we consider the simplest model, the linear approximation. Near the center of the nth syllable t_n, the linearized pitch contour is

$$p(t) = A_n + B_n(t - t_n), \qquad (8.24)$$

and similarly, around time t_{n+1},

$$p(t) = A_{n+1} + B_{n+1}(t - t_{n+1}). \qquad (8.25)$$

We need to construct a pitch curve in the interval $t_n < t < t_{n+1}$ such that in the neighborhood of t_n and t_{n+1}, the pitch value and its derivative approach the original values, but also connect continuously through the entire interval $t_n < t < t_{n+1}$. A polynomial of the lowest order satisfying such a condition is

$$p(t) = A_n + B_n(t - t_n) + C_n(t - t_n)^2 + D_n(t - t_n)^3, \qquad (8.26)$$

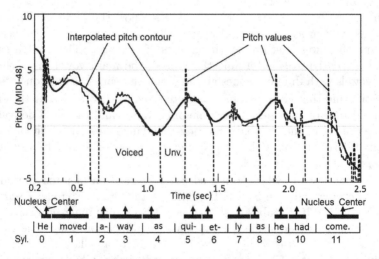

Fig. 8.19. Interpolated global pitch contour. Using the coefficients of the linear approximations of the pitch contour on each syllable, as well as higher-order polynomials, a continuous curve of the pitch contour of the entire sentence is generated [10].

with

$$C_n = \frac{3(A_{n+1} - A_n)}{\Delta t^2} - \frac{(2B_{n+1} + B_n)}{\Delta t}, \tag{8.27}$$

and

$$D_n = \frac{2(A_{n+1} - A_n)}{\Delta t^3} + \frac{(B_{n+1} + B_n)}{\Delta t^2}, \tag{8.28}$$

where

$$\Delta t = t_{n+1} - t_n. \tag{8.29}$$

The validity of the formulas can be checked directly. Using those interpolation formulas, the pitch values and their derivatives are continuous over the entire record.

Figure 8.19 shows the result. At the center of each syllable, the values of pitch and time derivative of pitch equal that of the linear approximation, and the curve is continuous up to the second derivative. From considering the interpolated pitch contour, it is clear that the contour follows closely the original pitch values, with short-time smoothing. During voice recovery, after jitter and shimmer are added, the recovered voice sounds natural, as shown by the samples.

The interpolated pitch contour makes no difference between voiced sections and unvoiced sections. This feature is consistent with the physiology of voice production and the parameterization of the speech signal. On one hand, during production of human voice, pitch is controlled by the muscles in the larynx. Even in unvoiced sections, these muscles continue to move. The conditions of pitch generation change continuously. On the other hand, the parameterization of the speech signal in terms of timbre vectors requires a continuous frame segmentation into the unvoiced sections.

8.5.2 Extraction of parameters

In this section, methods to extract expansion coefficients from the pitch values in each syllable nucleus are presented.

A glance into the pitch data from any speech records reveals that the pitch data are often not ideal. Especially at the beginning and at the end of a voiced section, the pitch values can deviate from the values at the center of the voiced section substantially. In order to obtain accurate expansion coefficients for the pitch contour on each syllable nucleus, least-square fitting methods with certain *weight functions* are required. An effective mathematical tool can be found in the *Gegenbauer polynomials* $C_n^{(\lambda)}(x)$ [1, 64], which are defined through their generation function as

$$\frac{1}{(1 - 2xt + t^2)^\lambda} = \sum_{n=0}^{\infty} C_n^{(\lambda)}(x) t^n, \tag{8.30}$$

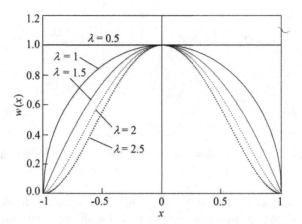

Fig. 8.20. Weight functions of Gegenbauer polynomials. Using least-squares fitting with a weight function, the inaccuracies of the pitch values at the two ends of each syllable can be mitigated [10]. For details see references [1, 64].

where $\lambda > -\frac{1}{2}$. The weight function, see Fig. 8.20, is

$$w(x) = \left(1 - x^2\right)^{\lambda - \frac{1}{2}}. \tag{8.31}$$

The Gegenbauer polynomials are orthogonal on [-1, 1] with the weight function, Eq. 8.31. For $n \neq m$,

$$\int_{-1}^{1} C_n^{(\lambda)}(x)\, C_m^{(\lambda)}(x) w(x)\, dx = 0, \tag{8.32}$$

and they are normalized by

$$\int_{-1}^{1} \left[C_n^{(\lambda)}(x) \right]^2 w(x)\, dx = \frac{\pi 2^{1-2\lambda}\, \Gamma(n + 2\lambda)}{n!\, (n + \lambda)\, [\Gamma(\lambda)]^2}. \tag{8.33}$$

The first three Gegenbauer polynomials are

$$C_0^{(\lambda)}(x) = 1, \tag{8.34}$$

$$C_1^{(\lambda)}(x) = 2\lambda x, \tag{8.35}$$

$$C_2^{(\lambda)}(x) = -\lambda + 2\lambda(1 + \lambda)x^2, \tag{8.36}$$

and

$$C_3^{(\lambda)}(x) = -2\lambda(1 + \lambda)\, x + \frac{4}{3}\lambda(1 + \lambda)(2 + \lambda)x^3. \tag{8.37}$$

The weight function determines the relative importance of the errors in the different parts of the interval of pitch data. As shown in Fig. 8.13, for $\lambda = 0.5$, the weight function is flat. The errors in the least-squares approximation is evenly distributed, and the Gegenbauer polynomials are Legendre polynomials.

For larger values of λ, the weight functions near the ends $x = -1$ and $x = 1$ are reduced, which means that the errors in the least-squares approximations near the ends are gradually neglected. The pitch data near the center of the syllables are more emphasized. For details, see Morse and Feshbach [64] or Abraham and Stegen [1]. Here, as an example, the case of $\lambda = 2$ is discussed in detail.

8.5.3 An exemplifying case

In this subsection, the procedure of obtaining the linear expansion coefficients of syllable pitch contours by application of Gegenbauer polynomials of order 2 is shown. The expressions of the polynomials are

$$C_0^{(2)}(x) = 1, \tag{8.38}$$

and

$$C_1^{(2)}(x) = 4x. \tag{8.39}$$

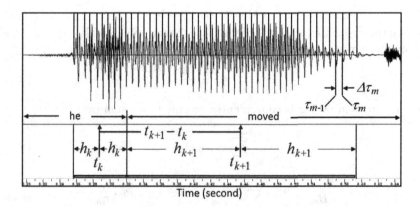

Fig. 8.21. Computation of expansion coefficients. The coefficients A_m and B_m of the linear approximation of the pitch contour on each syllable are computed by least-squares fitting with a weight function which deemphasizes the pitch values at the ends of the syllable, which are often less accurate than those near the center [10].

These polynomials are orthogonal over the interval $-1 < x < +1$, with a weight function $(1 - x^2)^{3/2}$ and normalization conditions

$$\int_{-1}^{1} \left[C_0^{(2)}(x) \right]^2 (1 - x^2)^{3/2} \, dx = \frac{3\pi}{8}, \tag{8.40}$$

$$\int_{-1}^{1} \left[C_1^{(2)}(x) \right]^2 (1 - x^2)^{3/2} \, dx = \pi. \tag{8.41}$$

Figure 8.21 show the details of the calculation process. For syllable k, the center is at $t = t_m$, and the half duration of the nucleus is h_m. For a pitch contour $p(t)$ in the time interval $[(t_m - h_m) < t < (t_m + h_m)]$, by a least-squares approximation using Gegenbauer polynomials of odrer 2 with $x = (t - t_m)/h_m$, the coefficients are

$$A_m = \frac{8}{3\pi} \int_{-1}^{1} p(t_m + h_m x) \, C_0^{(2)}(x) \, (1 - x^2)^{3/2} \, dx, \tag{8.42}$$

and

$$B_m = \frac{1}{\pi} \int_{-1}^{1} p(t_m + h_m x) \, C_1^{(2)}(x) \, (1 - x^2)^{3/2} \, dx. \tag{8.43}$$

The actual computation is based on a list of closure moments τ_m, where the pitch at period m is calculated from the difference of time between adjacent closure moments,

$$p_m = 69 - \frac{12}{\ln 2} \ln(440 \, \Delta\tau_m), \tag{8.44}$$

where

$$\Delta\tau_m = \tau_m - \tau_{m-1}. \tag{8.45}$$

Using the explicit expressions of the Gegenbauer polynomials, Eqs. 8.38 and 8.39, and noting that the parameter x at period m is

$$x_m = \frac{\tau_m - t_m}{h_m}, \tag{8.46}$$

the integration over x is conveniently executed as a sum over m:

$$A_m = \frac{8}{3\pi} \sum_{m:\, x_m > -1}^{x_m < 1} p_m \, (1 - x_m^2)^{3/2} \Delta\tau_m, \tag{8.47}$$

$$B_m = \frac{4}{\pi} \sum_{m:\, x_m > -1}^{x_m < 1} p_m \, x_m \, (1 - x_m^2)^{3/2} \, \Delta\tau_m. \tag{8.48}$$

The linear approximation of the pitch contour around $t = t_m$ is

$$p(t) \approx A_m + \frac{4B_m}{h_m}(t - t_m). \tag{8.49}$$

Fig. 8.22. Global pitch contours for various phrase types. The average global pitch contours of three types of phrases: declarative, interrogative, and intermediate phrases. The parameters for the pitch profiles are obtained by least-squares fit of pitch data with a polynomial of an appropriate order [10].

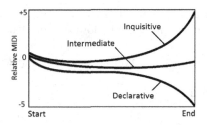

8.5.4 Global pitch contour and phrase type

By plotting the constant term of each syllable A_n against the index of syllables, the overall profile of the sentence or phrase is exhibited. The data points can be approximated by a polynomial using least-squares fit. Experiments showed that a fourth order polynomial is adequate,

$$p = c_0 + c_1 t + c_2 t^2 + c_3 t^3 + c_4 t^4, \tag{8.50}$$

where C_n are coefficients. By grouping sentences in a speech corpus into types, for example, declarative, interrogative, and intermediate phrases, and making an average for each type, a set of global pitch contour coefficients is established, see Fig. 8.22. For individual syllables, the constant term stored in the database is the relative value, that is, is the difference of the raw value and the global value.

During synthesis, according to the type of sentences or phrases, a global pitch contour is chosen, such as one of the pitch contour in Fig. 8.22. The constant term A_n for each syllable is the sum of the global pitch contour and the relative constant term in the database. The complete pitch contour curve is computed using the interpolation formula, Eq. 8.26.

8.5.5 Quadratic approximation

To include more details of the pitch contours, especially for tone languages, the pitch contours of each syllable can be expressed by a quadratic function with three coefficients A_m, B_m, and C_m. For a syllable centered at t_1, the pitch values can be approximated in a least-squares-fit sense as

$$p(t) = A_1 + B_1 t + C_1 t^2, \tag{8.51}$$

and similarly, around time t_2,

$$p(t) = A_2 + B_2 t + C_2 t^2. \tag{8.52}$$

We need to construct a pitch curve in the interval $t_1 < t < t_2$ such that in the neighborhood of t_1 and t_2, the values and derivatives are close to the original ones, yet the connection is continuous through the entire interval. A polynomial of the lowest order satisfying such a condition is

$$p(t) = Q_0 + Q_1 t + Q_2 t^2 + Q_3 t^3 + Q_4 t^4 + Q_5 t^5, \qquad (8.53)$$

with

$$Q_0 = A_1, \qquad (8.54)$$

$$Q_1 = B_1, \qquad (8.55)$$

$$Q_2 = C_1, \qquad (8.56)$$

$$Q_3 = \frac{10}{\Delta t^3}(A_2 - A_1) - \frac{2}{\Delta t^2}(3B_1 + 4B_2) + \frac{1}{\Delta t}(C_2 - 3C_1), \qquad (8.57)$$

$$Q_4 = -\frac{15}{\Delta t^4}(A_2 - A_1) + \frac{1}{\Delta t^3}(8B_1 + 7B_2) + \frac{1}{\Delta t^2}(3C_1 - 2C_2), \qquad (8.58)$$

and

$$Q_5 = \frac{6}{\Delta t^5}(A_2 - A_1) - \frac{3}{\Delta t^4}(B_1 + B_2) + \frac{1}{\Delta t^3}(C_2 - C_1), \qquad (8.59)$$

where $\Delta t \equiv (t_2 - t_1)$. The validity of these formulas can be tested by direct evaluation of the values at the center of the two syllables [10].

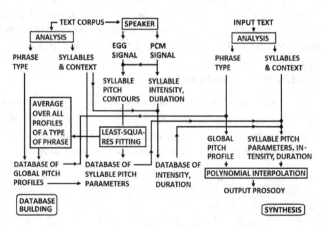

Fig. 8.23. Overall algorithm of prosody generation. The coefficients A_m and B_m of the linear approximation of the pitch contour on each syllable are computed by least-squares fitting with a weight function which deemphasizes the pitch values at the ends of the syllable, which are often less accurate than those near the center [10].

8.5.6 The general procedure

Figure 8.23 shows the process of building a database and the process of generating prosody during speech synthesis [10]. The left-hand side shows the database building process. A text corpus containing all the prosody phenomena of interest is compiled. A text analysis module segments the text into sentences and phrases, identifies the type of each sentence or phase of the text. The types comprise declarative, interrogative, imperative, exclamatory, intermediate phase, etc. Each sentence is then decomposed into syllables. Although automatic segmentation into syllables is possible, human inspection is often needed. The context information of each syllable is also gathered, comprising the stress level of the syllable in a word, the emphasis level of the word in the phrase, the part of speech and the grammatical identification of the word, and the context of the word with regard to neighboring words.

Every sentence in the text corpus is read by a professional speaker as the reference standard for prosody. The voice data are gathered through a microphone in the form of PCM (pulse-code modulation). If an electroglottograph instrument is available, the electroglottograph data (EGG) are simultaneously recorded. Both data sets are segmented into syllables to match the syllables in the text. Although automatic segmentation of the voice signals into syllables is possible, human inspection is often needed. From the PCM data, possibly combined with the EGG data through a segmentation procedure shown in Section 5.4, the pitch contour for each syllable is generated. Pitch is defined as a linear function of the logarithm of frequency or pitch period, preferably in MIDI units. The intensity and duration of each syllable are also identified from the PCM data.

The pitch contour in the voiced section of each syllable is approximated by a polynomial using least-squares fitting. The values of average pitch (the constant term of the polynomial expansion) of all syllables in a sentence or a phrase, are taken to form a polynomial using least-squares fitting. The coefficients are then averaged over all phrases or sentences of the same type in the text corpus to generate a global pitch profile for that type, see Section 8.5.4. The collection of those averaged coefficients of phrase pitch profiles, correlating to the phrase types, forms a database of global pitch profiles.

The pitch parameters of each syllable, after subtracting the value of global pitch profile at that time, are correlated with the syllable stress pattern and context information to form a database of syllable pitch parameters. That database will enable the generation of syllable pitch parameters by giving input information on syllables.

The right-hand side of Fig. 8.23 shows the process of generating prosody for an input text. First, by doing text analysis, the phrase type is deter-

mined. The phrase type – declarative, interrogative, exclamatory, intermediate phase, etc. – is determined, and a corresponding global pitch contour
is retrieved from the database. Then, for each syllable, the property and
context information of the syllable is generated. Based on that information, using the database, the polynomial expansion coefficients of the pitch
contour, as well as the intensity and duration of the syllable, are generated.
The global pitch contour is then added to the constant term of each set of
syllable pitch parameters. By using a polynomial interpolation procedure
as in Sections 8.5.1 and 8.5.5, an output prosody parameter set, including a continuous pitch contour for the entire sentence or phrase as well as
intensity and duration at every millisecond, is generated.

Appendix A
Kramers-Kronig Relations

Here is a proof of the Kramers-Kronig relations, or dispersion relations, Eq. 5.4, using Cauchy's theorem on contour integrals [2]. See Fig. A.1.

Let $\omega = x + iy$, Eq. 5.2 becomes

$$F(x + iy) = \frac{1}{2\pi} \int_0^\infty \xi(t)\, e^{ixt - yt}\, dt. \tag{A.1}$$

Because t is always positive, and $\xi(t)$ is finite, for large positive y, the value of $F(x + iy)$ vanishes exponentially with y. Consequently the logarithm of $F(x + iy)$ vanishes with y at least as fast as $1/y$.

Now look at an integral of the complex function

$$f(\omega') = \frac{\ln F(\omega')}{\omega' - \omega} = \frac{\ln A(\omega') + i\phi(\omega')}{\omega' - \omega} \tag{A.2}$$

on the contour described in Fig. A.1, which has no singularities. According to Cauchy's theorem, the contour integral, which includes an integral on the large circle with radius R, a line integral on the x-axis excluding a small section from $\omega - \epsilon$ to $\omega + \epsilon$, and a small circle of radius ϵ around ω, is zero.

$$\oint f(\omega')d\omega' = 0. \tag{A.3}$$

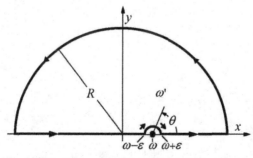

Fig. A.1.   Proof of dispersion relations for speech signals. Because of the causality condition of a timbron, the Cauchy contour integral is zero.

The components of the contour integral are evaluated as follows. The first one, the integral on the large circle, approaches zero as $R \to \infty$. The second component is an integral on the x-axis from $-\infty$ to $+\infty$, excluding a small interval $\omega' = \omega - \epsilon$ to $\omega' = \omega + \epsilon$. The third part is the integral on the half circle around the singularity. Let $\omega' - \omega = \epsilon e^{i\theta}$, as $\epsilon \to 0$, $f(\omega') \to f(\omega)$. The integral is

$$\int_\pi^0 d\left(\epsilon e^{i\theta}\right) \frac{\ln A(\omega) + i\phi(\omega)}{\epsilon e^{i\theta}} = -i\pi \left[\ln A(\omega) + i\phi(\omega)\right]. \qquad (A.4)$$

Separating the real part and the imaginary part of Eq. A.3, a pair of *dispersion relations* is obtained,

$$\phi(\omega) = -\frac{1}{\pi} \lim_{\epsilon \to 0} \left[\int_{-\infty}^{\omega-\epsilon} \frac{\ln A(\omega')}{\omega' - \omega} d\omega' + \int_{\omega+\epsilon}^\infty \frac{\ln A(\omega')}{\omega' - \omega} d\omega'\right], \qquad (A.5)$$

and

$$\ln A(\omega) = \frac{1}{\pi} \lim_{\epsilon \to 0} \left[\int_{-\infty}^{\omega-\epsilon} \frac{\phi(\omega')}{\omega' - \omega} d\omega' + \int_{\omega+\epsilon}^\infty \frac{\phi(\omega')}{\omega' - \omega} d\omega'\right]. \qquad (A.6)$$

Therefore, if the values of the amplitude of the Fourier transform of a casual function are known for all frequencies, the values if its phase for all frequencies can be calculated; and if the values of the phase of the Fourier transform of a casual function are known for all frequencies, the values of its amplitude for all frequencies can be calculated.

Appendix B
Laguerre Functions

Laguerre polynomials [1, 2, 64] are the eigenfunctions of weight function $x^k e^{-x}$ on the interval $(0, \infty)$, which can be defined by the Rodrigues formula

$$L_n^{(k)}(x) = \frac{e^x}{n! x^k} \frac{d^n}{dx^n} (e^{-x} x^{n+k}). \tag{B.1}$$

where n and k are non-negative integers. Explicitly, the first three Laguerre polynomials are

$$L_0^{(k)}(x) = 1, \tag{B.2}$$

$$L_1^{(k)}(x) = -x + k + 1, \tag{B.3}$$

and

$$L_2^{(k)}(x) = \frac{x^2}{2} - (k+2)x + \frac{(k+2)(k+1)}{2}. \tag{B.4}$$

In general,

$$L_n^{(k)}(x) = \sum_{m=0}^{n} (-1)^m \frac{(n+k)!}{(n-m)!(n+m)!m!} x^m. \tag{B.5}$$

Laguerre polynomials can be computed using the recurrence relation from those of lower orders

$$L_n^{(k)}(x) = \frac{2n+k-1-x}{n} L_{n-1}^{(k)}(x) - \frac{n+k-1}{n} L_{n-2}^{(k)}(x). \tag{B.6}$$

The Laguerre functions

$$\Phi_n^{(k)}(x) = \sqrt{\frac{n!}{(n+k)!}} \; e^{-x/2} x^{k/2} L_n^{(k)}(x) \tag{B.7}$$

are orthonormal on the interval $(0, \infty)$,

$$\int_0^\infty \Phi_m^{(k)}(x) \, \Phi_n^{(k)}(x) \, dx = \delta_{m,n}. \tag{B.8}$$

We have shown that Laguerre functions of order 4 are the simplest case for the expansion of amplitude spectra. For $k = 4$, the related formulas are:

$$L_0^{(4)}(x) = 1, \tag{B.9}$$

$$L_1^{(4)}(x) = -x + 5, \tag{B.10}$$

and

$$L_2^{(4)}(x) = \frac{x^2}{2} - 6x + 15. \tag{B.11}$$

The recurrence relation is

$$L_n^{(4)}(x) = \frac{2n + 3 - x}{n} L_{n-1}^{(4)}(x) - \frac{n + 3}{n} L_{n-2}^{(4)}(x). \tag{B.12}$$

The Laguerre functions are

$$\Phi_n^{(4)}(x) = \sqrt{\frac{n!}{(n + 4)!}} \; e^{-x/2} x^2 \, L_n^{(4)}(x). \tag{B.13}$$

Bibliography

[1] M. Abramowitz and I. Stegun. *Handbook of Mathematical Functions.* Dover Publications, New York, 1972.

[2] G. Alfken. *Mathematical Methods for Physicists.* Academic Press, New York, 1968.

[3] R. J. Baken. Electroglottography. *Journal of Voice*, 6:98–110, 1992.

[4] R. J. Baken. An overview of laryngeal function for voice production. *Professional Voice, Third Edition, Robert T. Sataloff, Plural Publishing*, 1:237–256, 2005.

[5] R. J. Baken and R. F. Orlikoff. *Clinical Measurement of Speech and Voice.* Singuler Publishing Group, San Diego, CA, 2000.

[6] J. Benesty, M. Mohan Sondhi, and Y. Huang. *Springer Handbook of Speech Processing.* Springer, 2008.

[7] G. Boehme and M. Gross. *Stroboscopy and Other Methods for the Analysis of Vocal Fold Vibration.* Whurr Publishers, London, 2005.

[8] R. Buderi. *Engines of Tomorrow, How the World's Best Companies Are Using Their Research Labs to Win the Future.* Simon and Schuster, New York, pages 150–151, 2000.

[9] F. Charpentier and M. Stella. Diphone synthesis using an overlap-add technique for speech waveform concatenation. *ICASSP'86*, 11:2015–2018, 1986.

[10] C. J. Chen. Prosody generation using syllable-centered polynomial representation of pitch contours. *U.S. Patent 8,886,539*, Nov. 11, 2014.

[11] C. J. Chen. System and method for speech synthesis. *U.S. Patent 8,719,030*, May 6, 2014.

[12] C. J. Chen. System and method for voice transformation. *U.S. Patent 8,744,854*, June 3, 2014.

[13] C. J. Chen. Pitch-synchronous speech coding based on timbre vectors. *U.S. Patent 9,135,923*, Sep. 15, 2015.

[14] C. J. Chen. System and method for speech recognition using pitch-synchronous spectral parameters. *U.S. Patent 8,942,977*, Jan. 27, 2015.

[15] C. J. Chen, R. A. Gopinath, M. D. Monkowski, and M. A. Picheny. Statistical acoustic processing method and apparatus for speech recognition using a toned phoneme system. *U.S. Patent 5,751,905*, May 12, 1998.

[16] C. J. Chen, R. A. Gopinath, M. D. Monkowski, and K. Shen. New methods in continuous mandarin speech recognition. *Eurospeech'97*, pages 1543–1546, 1997.

[17] C. J. Chen, K. F. Guo, P. Li, and L. Q. Shen. Method and apparatus for recognizing tone languages using pitch information. *U.S. Patent 6,510,410*, Jan. 21, 2003.

[18] C. J. Chen, H. Li, K. Shen, and G. Fu. Recognizing tone languages using pitch information on the main vowel of each syllable. *ICASSP'01*, 1:61–64, 2001.

[19] C. J. Chen, F. H. Liu, and M. A. Picheny. Automatic segmentation of continuous text using statistical approaches. *U.S. Patent 5,806,021*, Sep. 8, 1998.

[20] J. Clark and C. Yallop. *Phonetics and Phonology*. Blackwell, Oxford UK, 1990.

[21] B. Cranen and L. Boves. Pressure measurements during speech production using semiconductor miniature pressure transducers: Impact on models for speech production. *J. Acoust. Soc. Am.*, 77:1543–1551, 1985.

[22] D. Crystal. *Oxford Concise Dictionary of Linguistics*. Blackwell, Oxford UK, 1985.

[23] R. Daniloff, G. Schuckers, and L. Feth. *The Physiology of Speech and Hearing*. Prentice-Hall, Englewood Cliffs, NJ, 1980.

[24] S. B. Davis and P. Mermelstein. Comparison of parametric representations for monosyllabic word recognition in continuously spoken sentences. *IEEE Transaction on ASSP*, 28:357–366, 1980.

[25] L. Deng and D. O'Shaughnessy. *Speech Processing*. Marcel Dekker, 2003.

[26] L. Euler. Dissertation physica de sono, translated and annotated by ian bruce. *Euler Archive*, E002, 1727.

[27] P. Fabre. Un procédé électrique percutané d'inscription de l'acclement glottique au cours de la phonetion: glottographie de haute fréquence. premiers résultats. *Bulletin de l'Académie nationale de médecine*, 141:66–69, 1957.

[28] P. Fabre. La glottographie électrique en haute fréquence, particuralités de l'appareillage. *Comptes Rendus, Sociéte de Biologie*, 153:1361–1364, 1959.

[29] J. L. Flanagan and L. L. Landgraf. Self-oscillating source for vocal-tract synthesizers. *IEEE Transactions on Audio and Electroacoustics*, AU-16:57–64, 1968.

[30] M. Fujimoto. Pitch-synchronous speech coding by applying multiple analysis to select and align a plurality of types of code vectors. *U.S. Patent 5,864,797*, Jan. 26, 1999.

[31] S. A. Fulop. *Speech Spectrum Analysis*. Springer, Berlin, FRG, 2011.

[32] S. Furui. *Digital Speech Processing, Synthesis, and Recognition*. Marcel Dekker, New York, 2001.

[33] A. Gersho and R. M. Grey. *Vector Quantization and Signal Compression*. Kluwer Academic Publishers, Boston, 1992.

[34] S. Gold and N. Morgan. *Discrete-Time Speech Processing*. Prentice-Hall, Englewood Cliffs, NJ, 2002.

[35] L. Hanzo, F. C. A. Somerville, and J. P. Woodard. *Voice Compression and Communications*. IEEE Press, Piscataway, NJ, 2001.

[36] C. M. Harris. *Handbook of Acoustical Measurements and Noise Control*. McGraw-Hill, New York, 1991.

[37] F. J. Harris. On the use of windows for harmonic analysis with the discrete fourier transform. *Proceedings of the IEEE*, 66:51–83, 1978.

[38] H. L. F. Helmholtz. *On the Sensations of Tone, translated by H. Margenau into English*. Dover Publications, New York, 1954.

[39] I. P. Herman. *Physics of the Human Body*. Springer, Berlin, FRG, 2007.

[40] L. Hermann. Phonophotographische Untersuchungen I. *Pflueger's Archiv*, 45:582–592, 1889.

[41] L. Hermann. Bemerkungen zur Vokalfrage. *Pflueger's Archiv*, 48:181–194, 1890.

[42] L. Hermann. Phonophotographische Untersuchungen II und III. *Pflueger's Archiv*, 47:44–53, 347–391, 1890.

[43] L. Hermann. Beträge zur Lehre von der Klangwahrnehmung. *Pflueger's Archiv*, 56:467–499, 1894.

[44] W. J. Hess. A pitch-synchronous digital feature extraction system for phonemic recognition of speech. *IEEE Transactions on ASSP*, ASSP-24:14–25, 1976.

[45] R. J. Heuer, M. J. Hawkshaw, and R. T. Sataloff. The clinical voice laboratory. *Professional Voice, Third Edition, Robert T. Sataloff, Plural Publishing*, 1:355–394, 2005.

[46] M. Hirano. Morphological structure of the vocal cord as a vibrator and its variations. *Folia Phoniatr.*, 26:89–94, 1974.

[47] O. Hirschey and D. Klyve. The missing meditatio: Leonhard Euler's contribution to articulatory phonetics. *Historiographia Linguistica*, 42:63–83, 2015.

[48] K. Ishizaka and J. L. Flanagan. Synthesis of voiced sounds from a two-mass model of the vocal cords. *Bell Sys. Tech. J*, 51:1233–1268, 1972.

[49] F. Jelinek. *Statistical Methods for Speech Recognition*. The MIT Press, Cambridge MA, 1997.

[50] R. LeBlanc Jopling. Voice initiation and voice offset patterns in nomal females. *Thesis, Louisiana State University*, May 2009, 2009.

[51] J. Kadis. *The Science of Sound Recording*. Elsevier, New York, 2012.

[52] J. Kominek and A. Black. CMU ARCTIC Databases for Speech Synthesis. *CMU Language Technologies Institute, Tech report*, CMU-LTI-03-177, 2003.

[53] A. M. Kondoz. *Digital Speech: Coding for Low Bit Rate Communication Systems, Second Edition*. John Wiley and Sons Ltd, West Sussex, England, 2004.

[54] P. Ladefoged. *Elements of Acoustic Phonetics*. The University of Chicago Press, Chicago and London, 1966.

[55] P. Ladefoged and I. Maddieson. *The Sounds of the World's Languages*. Blackwell Publishers, Melden, MA, 1996.

[56] R. B. Lindsay. The story of acoustics. *J. Acoust. Soc. Am.*, 393:629, 1966.

[57] J. E. Miller M. V. Mathews and E. E. David Jr. Pitch synchronous analysis of voiced sounds. *J. Acoust. Soc. Am.*, 33:179, 1961.

[58] J. D. Markel and Jr. A. H. Gray. *Linear Prediction of Speech*. Springer, Berlin, FRG, 1976.

[59] P. H. Matthews. *A Dictionary of Linguistics and Phonetics*. Oxford University Press, Oxford UK, 1997.

[60] Y. Medan and E. Yair. Pitch synchronous spectral analysis scheme for voiced speech. *IEEE Trans. ASSP*, 37:1321–1328, 1989.

[61] D. Miller and H. Schutte. Characteristic patters of sub- and supra-glottal pressure variations within the glottal cycle. *Transcripts of the Thirteenth Sump. Care of the Professional Voice I: Scientific papers*, 1984:77–75, 1984.

[62] D. G. Miller. *Resonance in Singing*. Inside View Press, Princeton, NJ, 2008.

[63] P. M. Morse. *Vibration and Sound*. American Institute of Physics, 2007.

[64] P. M. Morse and H. Feshbach. *Methods of Theoretical Physicists*. McGraw-Hill, New York, 1953.

[65] E. Moulines and F. Charpentier. Pitch-synchronous waveform processing techniques for text-to-speech synthesis using diphones. *Speech Commun.*, 9:435–467, 1990.

[66] E. Moulines and J. Laroche. Techniques for pitch-scale and time-scale transformation of speech. *Speech Commun.*, 16:175–205, 1995.

[67] H. M. Nussenzveig. *Causality and Dispersion Relations*. Academic Press, NewYork, 1972.

[68] Handbook of the International Phonetic Association. *International Phonetic Association*. Cambridge University Press, 1999.

[69] T. F. Quatieri. *Speech and Audio Signal Processing.* John Wiley and Sons, 2000.

[70] L. R. Rabiner and R. W. Schafer. *Digital Processing of Speech Signals.* Prentice-Hall, Englewood Cliffs, NJ, 1978.

[71] L. R. Rabiner and R. W. Schafer. *Fundamentals of Speech Recognition.* Prentice-Hall, Englewood Cliffs, NJ, 1993.

[72] Lord Rayleigh. *The Theory of Sound.* Dover Publications, New York, 1945.

[73] G. O. Russell. *The Vowel.* McGrath Publishing Company, College Park, MD, 1928.

[74] R. T. Sataloff. The human voice. *Scientific American,* December 1992:108, 1992.

[75] R. T. Sataloff. Clinical anatomy and physiology of the voice. *Professional Voice, Third Edition, Robert T. Sataloff, Plural Publishing,* 1:143–178, 2005.

[76] R. T. Sataloff. Vocal Health. *www.otopa.org/patient/vocalhealth.shtml,* 2014.

[77] R. T. Sataloff, S. Mandel, R. Man non Espaillat, Y. D. Heman-Ackah, and M. M. Abeza. Laryngeal electromyography. *Professional Voice, Third Edition, Robert T. Sataloff, Plural Publishing,* 1:395–423, 2005.

[78] S. T. Sataloff. *Professional Voice, the Science and Art of Clinical Care, Third Edition.* Prural Publishing, San Diego, Oxford, 2005.

[79] S. T. Sataloff, S. Mandel, Y. Heman-Ackah, R. Ma non Espaillat, and M. Abaza. *Laryngeal Electromyography, Second Edition.* Prural Publishing, San Diego, Oxford, 2005.

[80] R. C. Scherer. Laryngeal function during phonetion. *Professional Voice, Third Edition, Robert T. Sataloff, Plural Publishing,* 1:257–274, 2005.

[81] J. Schroeter. *Basic Principles of Speech Synthesis.* Springer Handbook of Speech Processing, Springer, 2008.

[82] H. Schutte and D. Miller. The effect of F0/F1 coincidence in soprano high notes on pressure at the glottis. *J. Phonetics,* 14:385–392, 1986.

[83] E. W. Scripture. *The Elements of Experimental Phonetics*. Charles Scribner's Sons, New York, 1902.

[84] E. W. Scripture. Die Nature der Vokale. *Zeits. f. Exp. Phonetik*, 1:16–33, 115–146, 1930.

[85] D. H. Staelin and C. R. Cabrera-Mercader. Pitch-synchronous speech processing. *US Statutory Invention Registration*, US H2172 H, 2006.

[86] K. N. Stevens. *Acoustic Phonetics*. The MIT Press, Cambridge, MA, 2000.

[87] K. N. Stevens and M. Hirano. *Vocal Fold Physiology*. Univesity of Tokyo Press, Tokyo, 1981.

[88] B. H. Story and I. R. Titze. Voice simulation with a body-cover model of the vocal folds. *J. Acoust. Soc. Am.*, 97:1249–1260, 1995.

[89] C. Sturt, S. Villette, and A. M. Kondoz. LSF quantization for pitch synchronous speech coders. *ICASSP-2003*, 2:165–168, 2003.

[90] Y. Stylianou. Voice transformation. *Springer Handbook of Speech Processing*, pages 489–503, 2008.

[91] Y. Stylianou. Voice transformation: a servey. *ICASSP'09*, pages 3585–3588, 2009.

[92] J. Sundberg. *Science of the Singing Voice*. Northern Illinois University Press, Dekalb, IL, 1987.

[93] J. G. Svec and H. K. Schutte. Videokymography: High-speed line scanning of vocal fold vibration. *Journal of Voice*, 10:201–205, 1996.

[94] R. Taori, R. J. Sluijter, and E. Kathmann. Speech compression using pitch synchronous interpolation. *ICASSP'95*, 1:512–515, 1995.

[95] I. R. Titze. *Principles of Voice Prodution*. Prentice-Hall, Englewood Cliffs, NJ, 1994.

[96] K. Tokuda, Y. Nankaku, J. Yamagishi K. Tokuda, H. Zen, and K. Oura. Speech synthesis based on hidden Markov models. *Proceedings of the IEEE*, 101:1234–1252, 2009.

[97] J. S. Toll. Causality and the dispersion relation: Logical foundations. *Phys. Rev.*, 104:1760–1770, 1956.

[98] R. Wheatstone. Reed organ-pipes, speaking machines, etc. *Westminster Review*, 28:27, 1837.

[99] R. Willis. On the vowel sounds, and on reed organ pipes. *Camb. Phil. Trans.*, III:231, 1829.

[100] G. S. K. Wong and T. F. W. Embleton. *Handbook of Condenser Microphones*. American Institute of Physics, Woodbury, New York, 1995.

[101] H. Yang, S.-N. Koh, and P. Sivaprakasapillai. Pitch synchronous multi-band (PSMB) speech coding. *ICASSP'95*, 1:516–519, 1995.

[102] B. Yin and M. Felley. *Chinese Romanization: Pronunciation and Orthography*. Sinolingua, Beijing, 1990.

[103] M. Yip. *Tone*. Cambridge University Press, Cambridge, UK, 2002.

[104] H. Zen, K. Tokuda, and A. W. Black. Statistical parametric speech synthesis. *Speech Communication*, 51:1039–1064, 2009.

Index